Lecture Notes in Energy

Volume 86

Lecture Notes in Energy (LNE) is a series that reports on new developments in the study of energy: from science and engineering to the analysis of energy policy. The series' scope includes but is not limited to, renewable and green energy, nuclear, fossil fuels and carbon capture, energy systems, energy storage and harvesting, batteries and fuel cells, power systems, energy efficiency, energy in buildings, energy policy, as well as energy-related topics in economics, management and transportation. Books published in LNE are original and timely and bridge between advanced textbooks and the forefront of research. Readers of LNE include postgraduate students and non-specialist researchers wishing to gain an accessible introduction to a field of research as well as professionals and researchers with a need for an up-to-date reference book on a well-defined topic. The series publishes single- and multi-authored volumes as well as advanced textbooks.

Indexed in Scopus and EI Compendex The Springer Energy board welcomes your book proposal. Please get in touch with the series via Anthony Doyle, Executive Editor, Springer (anthony.doyle@springer.com)

More information about this series at https://link.springer.com/bookseries/8874

Oliver Inderwildi · Markus Kraft
Editors

Intelligent Decarbonisation

Can Artificial Intelligence and Cyber-Physical
Systems Help Achieve Climate Mitigation
Targets?

 Springer

Editors
Oliver Inderwildi ⓘ
Cambridge CARES
University of Cambridge
Singapore, Singapore

Markus Kraft ⓘ
Cambridge CARES
University of Cambridge
Singapore, Singapore

ISSN 2195-1284   ISSN 2195-1292   (electronic)
Lecture Notes in Energy
ISBN 978-3-030-86214-5   ISBN 978-3-030-86215-2   (eBook)
https://doi.org/10.1007/978-3-030-86215-2

This Springer imprint is published by the registered company Springer Nature Switzerland AG
The registered company address is: Gewerbestrasse 11, 6330 Cham, Switzerland

Oliver Inderwildi dedicates this book to his beloved late grandparents for their lifelong support of him and his career; they were there for him when no one else was.

Irma Agnes Inderwildi

21 January 1931 – 19 November 2021

Richard Albert Inderwildi

30 March 1932 – 1 October 2021

Preface

Singapore is a densely populated city-state on a low-lying island just one degree north of the equator. It is at particular risk from the undesirable effects of climate change, including sea level rise, variable weather and its impact on water resources and global food supplies, public health issues such as endemic vector-borne diseases, and the urban heat island effect. These are potentially existential threats that Singapore takes very seriously.

A large part of the carbon emissions that contribute to climate change originates from the combustion of fossil fuels for energy. The most direct way to reduce emissions would be to cut down the use of fossil fuels. However, Singapore has limited access to alternative or renewable energy, hence reducing demand with better energy efficiency becomes the focus of Singapore's efforts. Here science plays a critical role and just as Singapore has embraced and prioritised technological solutions that have allowed it to flourish and to become a smarter city, a scientific approach will be key to developing the technologies that are needed to decarbonise.

2021 marks the 30th year since the beginning of systematic public investment to develop research and development capabilities in science and technology. This public investment has seen the level of scientific research capability in Singapore rise to the level of similarly sized advanced economies around the world. Part of this rise can be attributed to the development of a unique international research collaboration model in CREATE (Campus for Research Excellence and Technological Enterprise). This model accelerates scientific progress in Singapore, gathering the world's best research institutions and universities to work together on problems that affect Singapore and the world, at a scale that has the potential to deliver impact. The co-location of these institutions with local universities gives rise to organic collaboration and innovation, allowing novel and interdisciplinary research to take center stage. Since it began in 2007, CREATE has made strides in many diverse fields, from urban science to smart materials. Several of the excellent contributions in this book are the result of research at CREATE.

The explosive growth in the use of digital technologies like artificial intelligence, machine learning and smart grids offers new and promising pathways to develop more energy-efficient systems. These technologies have great potential to contribute to decarbonisation through the discovery of new mitigation strategies, novel materials development and greater efficiency of our current processes (for example, chemical industry and transportation).

It is this potential that *Intelligent Decarbonisation* seeks to explore and promote. This book gathers expertise from a range of sources to put forward a vision of an interconnected world that utilises diverse knowledge, experience and innovation for a more efficient and sustainable future. There is a range of perspectives by authors who have roles ranging from those of the national policymaker to leaders of top research institutions. There is a unique format that illustrates the potential of digitalisation not just through scientific articles but also by including interviews with those who work in the key areas of decarbonisation and artificial intelligence.

There are, of course, the technical contributions including the editors' own articles on the digital world avatar model and the history of artificial intelligence as well as several contributions from CREATE. Views on how some of the challenges of cyber-security, legal and governance issues should be addressed form a useful complement to the technologies described.

There is an excellent geographical spread of perspectives too, with authors writing from Singapore to North America and in between, amply illustrating the abundance of research on climate change and on digital technologies from universities, industry and government.

I congratulate the editors of *Intelligent Decarbonisation* for bringing all these perspectives and ongoing research together in this comprehensive and accessible format.

Singapore   Professor Khiang-Wee LIM
31 March 2022   Executive Director of CREATE
                                                                        (Campus for Research Excellence and
                                                                        Technological Enterprise) in Singapore

Acknowledgements

The editors, Dr. Oliver Inderwildi & Professor Dr. Markus Kraft, gratefully acknowledge Louise Renwick and Dr. Andrew Breeson for proofreading and copy editing, Ada See for editorial support and Joy Haughton for helpful discussions.

The editors like to extend their gratitude to Anthony Doyle and Saranya Kalidoss of Springer Nature for their flexibility and continued support during the planning, preparation and production of this manuscript.

The editors are moreover grateful for the institutional support from the Campus for Research Excellence and Technological Enterprise (CREATE) Singapore as well as the Department of Chemical Engineering & Biotechnology and Churchill College, both University of Cambridge.

Contents

Part V   The Big Picture

Part VI   Conclusions

Part I
Introduction

Chapter 1
Introduction

Oliver Inderwildi and Markus Kraft

Abstract Humanity is at a crossroads as it is facing *two* existential risks: Firstly, climate change caused by anthropogenic greenhouse gas (GHG) emissions could lead to an uninhabitable atmosphere and hence, overcoming it is not optional; it is deciding for humanity's survival and hence existential. Secondly, the advent of human-like or general artificial intelligence (AGI) which could—analogue to uncontrolled climatic change—lead to the extinction of humankind as it would be subject to the goodwill of a higher intelligence. Fortunately, both risks also present unprecedented opportunities: On the one hand, the restructuring the global economy to net-zero emissions would mitigate and reverse climate change and inevitably lead to a healthier, more liveable world. AI, on the other hand, could be utilised to address global problems the human mind is not capable of solving. This book brings together leading thinkers from all walks of life to discuss how AI and digital technologies could support the so-called Race to Zero and mitigate or even reverse climate change.

1.1 Introduction

In the US television series "Person of Interest" a supercomputer referred to as "The Machine" assists human protagonists in safeguarding their community. "The Machine" is an artificial general intelligence (AGI) initially built by the U.S. Government to predict and prevent imminent terrorist attacks by analysing immense amounts of surveillance data. Later it assists one of its architects—whom it refers to as Father—in preventing crime. In the end, this benevolent AGI prevents numerous terrorist attacks, capital and other major crimes while fending off a competing, malicious AGI. The Machine analyses potential threats by gathering information from surveillance cameras and online resources such as criminal data bases, thereby anticipating possible outcomes with probabilities. In case the probability for harm in any respect is high, the machine intelligence alerts its human collaborators to step in – the fictional AI's goal is to save human lives and avert harmful events and is truly benevolent.

O. Inderwildi (✉) · M. Kraft
Cambridge CARES, Cambridge, England
e-mail: oliver.inderwildi@essec.edu

© Springer Nature Switzerland AG 2022
O. Inderwildi and M. Kraft (eds.), *Intelligent Decarbonisation*, Lecture Notes
in Energy 86, https://doi.org/10.1007/978-3-030-86215-2_1

Fig. 1.1 The fictional AGI "The Machine" anticipating a probable violent crime based on extensive data analysis and probability projections combined with video surveillance

Figure 1.1 illustrates the workings of the AI imagined by director Jonathan Nolan: perpetrator and victim are anticipated with probabilities (94.72% and 92.71% resp.) and based on these high probabilities the machine intelligence concludes that violence is imminent. At present, such a type of intelligence appears to be true fiction, but we will see in Chap. 3 that advances in AI, big data and quantum computing might soon turn fiction into reality.

Literature, especially science fiction, is full of examples of malicious AI, while examples of benevolent AI—such as the one described above—are significantly rarer. The critical question is: What could be achieved if such an AI or even AGI could support us in all areas of life from health to wellbeing and from sustainability to wealth creation? In any case, as we will see in Chap. 4", AGI is still a relatively far-off concept, while AI's impacts can already be seen and measured. We will discuss questions regarding AI, including the existential risk it might pose, in Chap. 3. The main idea of this book is to assess how AI and cyber-physical systems (CPS)—associated, orchestrated digital technologies—can help humankind to overcome its most complex and most pressing challenge: climate change.

1.2   Climate Change

The emissions of greenhouse gases (GHG) affect the earth's climate and its habitability; since the industrial revolution starting in the North of England in the eighteenth century, the global economic system has cumulatively emitted more than

16 teratons of GHGs which have significantly affected our climate (Intergovernmental Panel on Climate Change 2014). According to The Intergovernmental Panel on Climate Change (IPCC), anthropogenic—or human-induced—warming reached 1 °C ± 0.2 °C above pre-industrial levels in 2017 (Intergovernmental Panel on Climate Change 2014). Global warming is defined by the IPCC and in its reports as an increase in combined surface, air and sea surface temperatures averaged over the globe and over a 30-year period.

A clear visualisation of the complexity of the problem that climatic change represents was published by the British researcher Ed Hawkins in 2018 (Hawkins 2021). The representation is referred to as the Warming Stripes; blue lines represent years with below average temperatures while red stripes represent years with above average temperatures. The graph therefore emphatically shows that the earth is heating up quickly with enormous ramifications for its habitability, water and food security, and the probability of extreme weather events, among many other issues (Fig. 1.2).

The probability of extreme weather events and the consequent implications for the habitability of the planet are hence intrinsically linked to the global average temperature, which in turn depends on the amount of GHG emitted (Inderwildi and King 2012). Owing to this existential question, the United Nations Framework Convention on Climate Change (UNFCC) has initiated the Paris Agreement. This aims to restrict global warming to 1.5 °C, which would substantially reduce the risks and impacts of climate change. In order to achieve this, global GHG emissions will have to be curbed drastically (United Nations Framework Convention on Climate Change 2015). The problem is that economic activity and GHG emissions are also intrinsically linked. The 2020–2021 Covid-19 crisis provides an illustration of this: The Financial Times reported on 14th of October 2020 that due to Covid-19 restrictions,

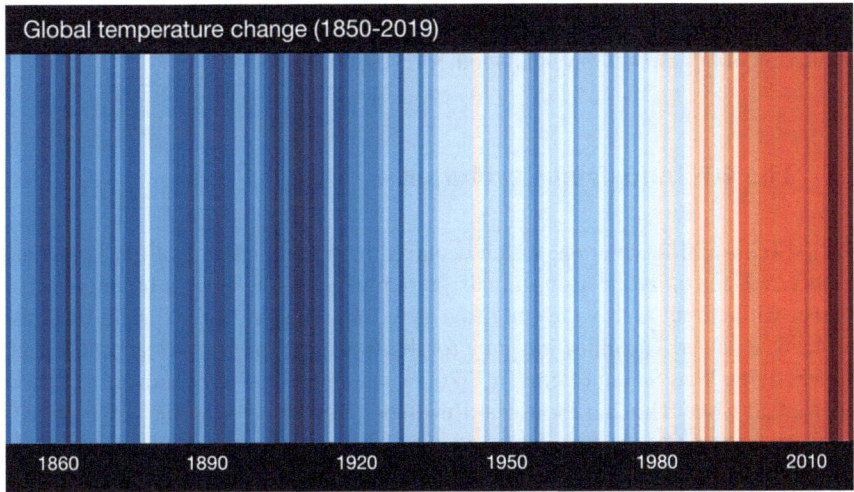

Global temperature change (1850-2019)

1860   1890   1920   1950   1980   2010

Fig. 1.2 The Warming Stripes by Ed Hawkins, a simple visualisation of the progress of climate change between 1850 and 2020 (red stripes denote above average temperatures, blue below average)

global average emissions fell by 8.8% in the first half of 2020 while economic output fell by approximately the same percentage (Hook 2020). Leading economies such as the UK, Germany and Japan, as well as innovation hubs such as Singapore and Switzerland, have adopted so-called net-zero policies that aim to fully decarbonise their economies by 2050.

Overcoming climate change is not optional, it is deciding for humanity's survival and hence this challenge is existential. Restructuring the global economy towards net-zero emissions would mitigate and reverse climate change and lead to a healthier, more liveable world.

The key idea of this book is to assess how state-of-the-art digital technologies—ranging from basic digital processes to advanced AI—are affecting and will affect the emission mitigation and hence ameliorate the negative impacts of climate change. Part II of this compendium will introduce the oncoming technologies that are critical for emission reduction starting with cyber-physical systems or CPS (Chap. 2, Zhang et al.), artificial intelligence (Chap. 3, Inderwildi and Kraft), dynamic knowledge graphs (Chap. 4, Lim et al.) as well as blockchain and smart contracts (Chap. 7, Bin et al.). Part III will assess specific sectors ranging from the producing industries to water provisions and the management of urban systems. Part IV will provide legal scholars and practitioners with the opportunity to voice their concern about the risks of the current and upcoming transformation, while proposing intelligent solutions that mitigate those risks. The chapters of all five parts of this book are complemented with interviews with leading practitioners and decision makers to create a holistic picture of the transformation, especially regarding the feasibility of the proposed digital solutions. The compendium will then conclude with the synthesis of the learnings and the common threads that are established and conclude with the pivotal take-away messages.

Is this transformation a narrative about the future or is it already ongoing? While the digital transformation has disrupted many economic sectors already, for industrial production, it has already brought another revolution.

1.3   The 4th Industrial Revolution

The 1st Industrial Revolution, which occurred in the second part of the eighteenth century, reduced the reliance on animals, human effort and biomass as primary energy sources while fossil fuels were embraced as the new energy source transformed into mechanical power. The inflection point for this transformation was clearly the development of the Watt Steam Engine in 1776, by which coal could be utilised to deliver mechanical work. The 2nd Industrial Revolution, also known as the Technological Revolution, started approximately a century later and spanned well into the early part of the twentieth century. A key breakthrough that initiated this revolution was the ability to harness electricity which led to new possibilities for energy provision, wired and wireless communication and many other advances. The Third Industrial

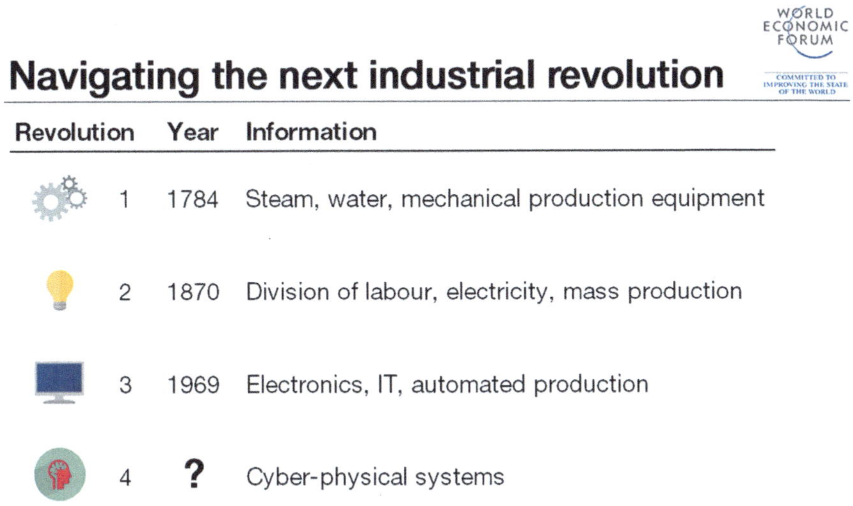

Navigating the next industrial revolution

Revolution	Year	Information
1	1784	Steam, water, mechanical production equipment
2	1870	Division of labour, electricity, mass production
3	1969	Electronics, IT, automated production
4	?	Cyber-physical systems

Fig. 1.3 The succession of industrial revolutions as defined by the World Economic Forum

Revolution began in the 1950s with the development of digital systems, communication and rapid advances in computing power, which have enabled new ways of generating, processing and sharing information. The dawn of digital systems laid the foundation for the advent of the 4th Industrial Revolution (Davis 2016; Schwab 2017) (Fig. 1.3). The key enabler of this revolution is the development of cyber-physical systems (CPS) which provide entirely new capabilities for people and machines by orchestrating both established and new capabilities (Inderwildi et al. 2020). As for the previous revolutions, the 4th Industrial Revolution relies on the technologies and infrastructure of the 3rd Industrial Revolution and consequently, progress seen over the last 200 years is clearly an evolution of technology that has successively altered the economic system and the functioning of society as a whole. The 4th Industrial Revolution goes well beyond the restructuring of industrial production as CPS have already entered our daily lives: smart homes, the streaming of media, online navigation and digital commerce are only a few examples of our daily use of CPS. The restructuring of industry itself, driven by digitalisation and CPS, is commonly referred to as Industry 4.0. Chapter 9 with Uwe Liebelt of BASF, Chap. 11 with Eckard Eberle of Siemens and Chap. 16 with Calvin Chung of JTC in Singapore will illustrate the enormous potential of this approach for producing industries from the practitioners' point of view. Chapter 8 by Sebastian Thiede of the University of Twente will specifically look at so-called cyber-physical production systems (CPPS) and will assess their current and future benefits. Oliver Lade of Capgemini Invent dives into the concept of Industry 4.0 for the chemicals industry, proposing and recommending a green, lean and digital approach for the chemicals industry.

1.3.1 Accelerating Change

Assessing state-of-the-art projections with the benefit of hindsight often reveals their inaccuracies, mainly because they are too conservative and neglect the rate of change and feedback loops between different technological development. For instance, the productivity of factories increases because of intrinsic improvements in their operations, but advances in robotics, data analysis, supply chain management etc. improve those intrinsic improvements. The higher productivity of factories then facilitates improvements in robotics and so on, creating feedback loops or indeed virtuous circles. To illustrate this, let's examine the evolution of levelised cost of energy (LCOE) for solar power over the last decade (IRENA 2019) and compare it to projections from 2010 (International Energy Agency 2010) (Fig. 1.4).

From this visualisation, Fig. 1.4, it becomes apparent that a decade ago improvements in this technology were clearly drastically underestimated. The true value for the LCOE of solar power provision (green line) in 2021 lies 82% below the 2010 estimate (blue line). Again, the likely source of this is the Industry 4.0 approach; advances in e.g. thin film technology and advanced manufacturing created synergetic feedback loops that accelerated improvements beyond what was believed to be possible in 2010—*the 4th Industrial Revolution is the inflection point for the transition to a net-zero economy.* Even more impressive is that solar power is now the cheapest way of providing energy that has ever existed. The LCOE of fossil-fuel power provision is shown as a grey band in Fig. 1.4 and as of 2020, solar power has undercut even the cheapest fossil energy provision (IEA 2020). Nevertheless, due to the intermittency of solar power challenges remain: the management of the electricity grid and the allocation of storage have to be enhanced while the grid of the future must be planned in such a way that it can cope with this intermittency. Chapter 12 by Baumgartner and Ulbig will propose that digital technologies not

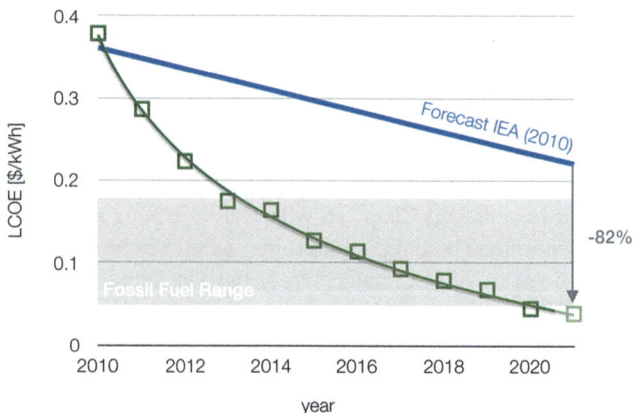

Fig. 1.4 The evolution of levelised cost of energy (LCOE) for solar power over the last decade (IRENA 2019) and compare it to projections from 2010

only assist with the management of the existing grid, but will help to plan the evolution to a more resilient grid; this will impact not only emissions but also operational and capital expenditure. The impact of digital technologies is not restricted to the provision of energy; it affects all areas of life.

1.3.2 Digitalisation and Digital Twins

The transformation of daily life through digitalisation can be seen in almost all areas: music and TV are streamed, travel tickets are bought via a mobile app, newspapers and magazines are read on tablets and payments are processed via electronic devices rather than by cash or card. These digitalisation trends reduce emissions by saving resources such as paper and plastic, while also increasing emissions by demanding electricity and precious resources such as lithium and cobalt. It is however anticipated that overall, the replacement of material services reduces emissions. Instead of driving somewhere to complete an administrative process, completing this task online will clearly reduce net emissions. As more and more proceedings of human affairs end up in the digital cloud, from social media posts to traffic situations to ticket purchases, a digital representation of the physical world is gradually formed—this virtual representation is referred to as the Digital Twin of the physical world. Bishop of Arup sets out in Chap. 18 how Digital Twins, equipped with models for travel demand and transport supply, can guide consumers onto not only the fastest journey from A to B, but also the one with the least environmental impact. In this chapter, the economist explains how AI will be helpful in this dynamic process by interlinking huge data sets and thus being able to make accurate predictions.

However, digital technologies are not restricted to the optimisation of the engineering aspects of energy provision. Hamacher, from the Technical University of Munich, proposes an Open Data Platform that—through information transparency—provides all market participants with the same information. The beneficial effects of such a symmetric market for the transformation of energy markets will be laid out in Chap. 19. Digital technologies not only impact specific areas responsible for GHG emission, but also the synchronisation of different energy systems i.e. heat and electricity. Such an intelligent synchronisation is of paramount importance for the deep decarbonisation of the economic system. Shah and co-workers from Imperial College London will lay out four distinct prerequisites for the successful actuation of emission and cost reduction by digital frameworks, including the recent technology leaps that have made these possible. The authors then use hands-on examples from energy provision, the built environment and coordination of urban districts to validate their hypotheses on digital technologies (Chap. 15).

However, not all urban systems are the same and local effects have to be considered and included as well. An illustrative example of such local effects are so-called Urban Heat Islands (UHI). Temperatures in urban, especially metropolitan, areas are usually elevated compared to the rural surroundings. These UHI are due to the configuration and increased mass of urban infrastructure, as well as anthropogenic waste heat and hot exhaust gases from heating, cooling, industry or transport; electrification of transportation, for instance, could drastically reduce waste heat as well as hot exhaust gases and thereby alleviate UHI in cities. This is especially relevant for cities in (sub-)tropical areas because of their intrinsic need for cooling; Schmitt and co-workers from the Singapore-ETH Centre will lay out in Chap. 17 how a Digital Urban Climate Twin (DUCT) can assist with the transition to net zero and counteract the formation of UHI.

Local effects are especially critical when considering the most important resource for human welfare and survival, water. As ground- and freshwater is becoming scarcer, desalination of sea water is becoming more critical for water security. As desalination of seawater is energy intense, the energy demand from water provision is set to increase significantly from the current 4% of global electricity demand. Beiji and Lade of Siemens, however, outline in Chap. 13 how intelligent digital technologies can counteract this and even lead to significant reductions with regard to the energy and carbon footprint of water provision. The authors define four critical factors in the water cycle responsible for energy demand and hence GHG emissions and showcase how digital technologies can actively reduce this demand with limited capital expenditure. The deductions on urban systems providing energy and water go hand-in-hand with the experience of the CEO of a typical German utility provider, Christoph Dörr, who sees digitalisation as a pivotal milestone in the integrated management of urban services from heat and electricity to potable water provision. JTC's Calvin Chung (Chap. 16) sets out how digital technologies facilitate not only the management of commercial and residential building complexes, but also the planning, design and building of these.

1.4 Risks and Opportunities of Artificial Intelligence

1.4.1 AI for Interconnecting and Improving Systems

Artificial intelligence already has an impact on our everyday lives. Suggestions from Netflix or Amazon are based on deep learning, drug discoveries have been enhanced and accelerated using computer intelligence while in hospitals, diseases are spotted in medical images that are far beyond the accuracy of the human eye (Colback 2020). The recent developments in AI have also been successfully implemented in energy efficiency improvements. DeepMind has developed an AI system—based on deep neural networks—which was used for cooling control in Google's data centers. Figure 1.5 illustrates the effectiveness of the deep learning algorithm: within a year, using fewer than 90 million training cycles, the computer intelligence was able to reduce energy consumption of data centres by a whopping 30% over traditional engineering (Inderwildi et al. 2020)! This was achieved because the machine intelligence chose unconventional starting points as well as actions and then observed whether its innovative steps led to an improvement or a deterioration. Because the AI learns from

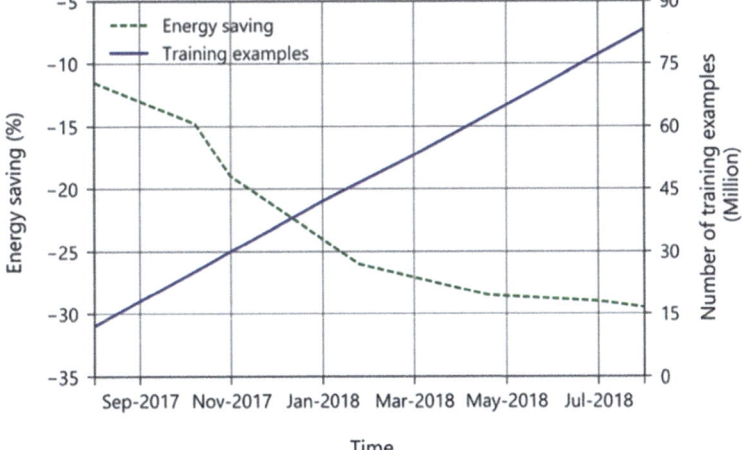

Fig. 1.5 Performance of DeepMind's AI system in improving the cooling of a data centre using deep neural networks. Energy saving (green) and number of training examples (blue) in a deep neural network (Inderwildi et al. 2020)

every improvement or indeed deterioration, future efficiency gains are highly likely. The described 30% decreases in energy consumption equate to a 30% reduction in emissions from energy. However, what could be achieved if simultaneously, the provision of energy could be improved by an analogue algorithm? The two efficiency improvements would reinforce one another and therefore truly aid the mitigation of GHG emissions and hence climate change.

The following chapters will first assess exactly this: what role in emission reduction can digital tools such as AI, CPS and blockchain play? Subsequently we shall look at specific sectors such as energy, transport or the urban environment, presenting examples to illustrate the ever-increasing speed of technological development.

1.4.2   The Risks of Machine Intelligence

From the example on data servers outlined above, it becomes clear that machine intelligence is ruthlessly efficient at achieving its goals. Unfortunately, this ruthlessness—which stems from the absence of a moral compass—generates yet another existential risk. What would happen if an AGI decides that the best way to protect the earth and its atmosphere is to adjust the human population to numbers that Mother Earth can sustain and what if this number is below the current world population? There is consensus in the scientific community that the first superintelligence is the last invention humankind has to create. The question is, what will this intelligence do with us, given it is unlikely to have a moral compass? The Swedish philosopher Nick Bostrom compares such a scenario to the situation of mountain gorillas: their fate depends on human goodwill and so might the fate of humanity depend on the actions of a future machine superintelligence (Bostrom 2014). According to Max Tegmark, society should not get overwhelmed by successes made with AI because success breeds ambition and this ambition might hurt us in the long run (Tegmark 2017). In any case, it has to be ensured that AI's goals are and stay aligned with humanity's goals; in Chap. 3, we will examine AGI, technological singularity and run-away scenarios that might affect mankind. Owing to these risks posed by AI and AGI as well as the potential for surveillance and privacy infringement introduced by CPS, Part IV will host legal scholars and governance specialists that address these issues and potential solutions. One of the key questions that must be solved is, how should intelligent, decision-making machines be addressed legally? Neupert, working for Kümmerlein Lawyers & Notaries, reviews the scenario that underlies the legal discussions: could machines imitate human thinking or reach a level that is not exclusively mechanical? The fine line between mere complex technical processes and sufficient autonomy warrants attention from the legal system and is hence addressed by the German lawyer and legal scholar in detail in Chap. 22. Cliff-Tavor of Oliver Wyman dives into the challenges for corporates that are posed by AI-enhanced cyber-attacks and how this impacts their cyber security. He also proposes that a Smart City approach needs special attention as it has to be secured in order to live up to its potential—useful surveillance can easily tip over into a digital dystopia. Going beyond purely legal aspects, Jelinek of the Taihe Institute and the OECD calls for international as well as public–private cooperation to tackle the AI challenge. In his chapter, the German scholar addresses the governance gap created by advances in AI and calls for global coordination to balance the technological and political dimensions of AI governance so that developments remain human- and environment-focused. In this chapter, the governance obstacles that impede the implementation of responsible AI are discussed in detail.

It is very likely that technological developments outpace the adaptation of the legal and governance framework due to their inertia already. How will it be possible to ensure safety and privacy in such a scenario? Viskovich of the Swiss Cyber Forum proposes technology-agnostic and adaptive laws and regulations that can keep up with technological developments. Moreover, the Australian attorney proposes an "ethics-by-design" framework that ensures that these digital technologies are aligned with humanity's core ethical values.

In Part II and III, academic and industrial researchers and scholars as well as managers discuss the potential and risks of these technologies. What remains to be answered is how institutions can prepare themselves for the transition and whether they see themselves as active players in it. Therefore, the bigger picture is considered in Part V; this section features government advisors and officials, heads of universities and research councils as well as leaders of specific initiatives and provides a holistic view on how institutions deal with these challenges while assessing in which areas they can play a key role to make the transition beneficial and safe for the global society and planet Earth. The portfolio of opportunities ranges from targeted graduate courses to retraining the workforce, and the establishment of research centres and initiatives aimed at solving the key obstacles.

1.5   Concluding Remarks

This introduction sets out the opportunities that digitalisation and artificial intelligence provide and proposes that it is already foreseeable that these instruments are and will be pivotal to integrate systems that either provide or supply energy. As decarbonisation is directly or indirectly linked to the demand and consequently the supply of energy, these tools are of paramount importance for the mitigation of climate change. Part II will give experts the opportunity to set out different technologies and their potential impacts, while Part III will pass the baton to industrial and academic researchers to present case studies from their industries and assess the emission reduction potential of both CPS and AI. All these benefits, however, come with distinct risks and Part IV will address the solutions provided by governance and legal systems to mitigate these risks.

This compendium will set out why the 4th Industrial Revolution is the inflection point that provides the opportunity to decarbonise the globalised economy and consequently reach net-zero emissions. One point, however, is clear: these technological developments have to serve humanity and the planet and cannot jeopardise the future of either!

References

Bostrom N (2014) Super-intelligence: paths, dangers, strategies. Oxford University Press, Oxford, UK

Colback L (2020) The impact of AI on business and society. The Financial Times 16th of October 2020. https://www.ft.com/content/e082b01d-fbd6-4ea5-a0d2-05bc5ad7176c

Davis N (2016) What is the 4th industrial revolution. World Economic Forum, Geneva, Switzerland

Hawkins E (2021) University of Reading & UK Met Office. https://showyourstripes.info

Hook L (2020) Global emissions fell 8.8% in the first half of 2020, study shows. The Financial Times 14.10.2020. https://www.ft.com/content/8715342f-8ad6-4fa6-a5c0-709d299f925e

IEA (2020) World energy outlook 2010. Paris, France

Inderwildi OR, Zhang C, Wang X, Kraft M (2020) The impact of cyber-physical systems on the decarbonization of energy. Energy Environ Sci 3(13):744–771

Inderwildi OR, King DA (2012) Energy, transport and the environment. Springer, London, UK

Intergovernmental Panel on Climate Change (2014) AR5 synthesis report: climate change 2014. Geneva, Switzerland

International Energy Agency (2010) World energy outlook 2010. France, Paris

IRENA (2019) Renewable power generation cost 2019. International Renewable Energy Agency, Abu Dhabi, United Arab Emirates. https://www.irena.org/-/media/Files/IRENA/Agency/Public ation/2020/Jun/IRENA_Power_Generation_Costs_2019.pdf

Schwab K (2017) The 4th industrial revolution. Portfolio Penguin, London, UK

Tegmark M (2017) Life 3.0: being human in the age of artificial intelligence. Knopf, New York City, USA

United Nations Framework Convention on Climate Change (2015) Adoption of the Paris agreement, 21st conference of the parties. United Nations, Paris

Part II
Methods & Technology

Chapter 2
Cyber-Physical Systems in Decarbonisation

Oliver Inderwildi, Chuan Zhang, and Markus Kraft

Digital technologies such as advanced metering, big data, machine learning, and the Internet of Things are receiving significant attention as they provide the potential to facilitate the decarbonisation of industry while requiring limited investments. The orchestration of these novel technologies, so-called cyber-physical systems (CPS), provides additional synergetic effects that increase efficiency of energy provision and industrial production, thereby optimising economic feasibility and environmental impact. This chapter assesses the current as well as the potential impact of digital technologies within CPS on decarbonisation endeavours. *Ad-hoc* calculation for selected applications of CPS and its subsystems estimates not only the economic

O. Inderwildi (✉)
CARES, Cambridge Centre for Advanced Research and Education in Singapore, 1 Create Way, CREATE Tower, #05-05, Singapore 138602, Singapore
e-mail: oliver.inderwildi@essec.edu

Swiss Academies of Sciences, Berne, Switzerland

C. Zhang (✉)
Andlinger Center for Energy and the Environment, Princeton University, 86 Olden Street, Princeton, NJ 08544, USA
e-mail: cz13@princeton.edu; chuanzhang@princeton.edu

M. Kraft
Department of Chemical Engineering and Biotechnology, University of Cambridge, Philippa Fawcett Drive, Cambridge CB3 0AS, UK
e-mail: mk306@cam.ac.uk

CARES, Cambridge Centre for Advanced Research and Education in Singapore, 1 Create Way, CREATE Tower, #05-05, Singapore 138602, Singapore

School of Chemical and Biomedical Engineering, Nanyang Technological University, 62 Nanyang Drive, Singapore 637459, Singapore

© Springer Nature Switzerland AG 2022
O. Inderwildi and M. Kraft (eds.), *Intelligent Decarbonisation*, Lecture Notes in Energy 86, https://doi.org/10.1007/978-3-030-86215-2_2

impact but also the emission reduction potential and illustrates that CPS alter the marginal abatement cost curve (MACC) while creating novel pathways for the transition to a low-carbon energy system.

2.1 Introduction

The main objective of this chapter is conceptualising the different developments of digital technologies and their impact on the emission reduction of technological processes and its economic feasibility (International Energy Agency 2017). The beauty of digital technologies is that they provide the potential to facilitate the decarbonisation process while requiring limited investments. Moreover, there is the option to retrofit existing technological infrastructure and the possibility for intelligent planning of the evolution of this infrastructure. A list of critical digital technologies and related application examples are listed hereafter:

Big Data: Big data refers to a new kind of science that analyzes or systematically extract information from data sets that are too large or complex to be dealt with by traditional data-processing applications. These novel techniques are especially beneficial for data sets that are dynamic—e.g. real-time data—because they can provide essential insights that facilitate the optimisation of current systems and aid the planning of the evolution of complex systems such as energy, industry or mobility. Big data applications will especially be discussed in Chaps. 4, 12 and 18.

Machine Learning (ML): Machine learning describes a branch of computer algorithms that have the capability to improve automatically through experience. ML is considered a precursor or prerequisite for artificial intelligence. ML applications are a means in their own right as they are critical for the enhancement of current systems management for instance in energy (see Chap. 12), transport, or urban systems. More on the fundamentals of the technology will be presented in the subsequent chapter.

Internet of Things: Network of connected devices that has the ability to collect information about the real world remotely and share it with other systems and/or devices through Machine-to-Machine communication is referred to as the Internet of Things (IoT) (Törngren et al. 2014; Gubbi et al. 2013). IoT enabled appliances are for instance present in smart homes (Samuel 2016) and are a key component of Industry 4.0. Chapters 5, 11 and 12 will address this.

Edge Computing: A critical enabler of the IoT is edge computing, in this approach CPUs are added where real-time processing of data is required, most often in the device itself and thereby form a distributed computing network. Localised computational capacity enables rapid responses and simultaneously saves bandwidth. From Chaps. 4, 15 and 18 it will become apparent that this technology is key to the establishment of eco-industrial parks, smart cities, intelligent mobility and many other sectors.

Advanced Metering Infrastructure: Such an infrastructure is an integrated system of smart meters, communications networks and data management systems that enables two-way communication between providers and customers. Chapters 4 and 8 will clarify why such an infrastructure is key to increasing efficiency.

Cloud Computing: Cloud computing is the delivery of computing services via the Internet, i.e. the cloud, which offers flexible resources and economies of scale. Users pay only for cloud services used lowering operational expenditure (OpEx), while simultaneously minimizing infrastructure needs and hence capital expenditure (CapEx), will provide examples of how cloud computing can reduce costs and reduce emissions.

Blockchain: A non-centralised digital transaction ledger that is public (National Grid System Operator UK 2018) and can facilitate as well as automate processes. This efficiency increase can translate directly into costs and emissions savings as Wang and co-workers will explain in detail in Chap. 7 and Lim et al.: will substantiate this in Chap. 4.

Smart Contracts: A smart contract is a computer protocol intended to digitally facilitate, verify, or enforce the negotiation or performance of a contract. Chapters 4 and 7 will explain how this facilitates the reduction of emissions, while Neupert of Kümmerlein Lawyers & Notaries will provide legal and philosophical notations on smart contracts in Chap. 22

Semantic Web: This development of the World Wide Web (Berners-Lee et al. 2001) structures data in web pages and tags them in such a way that it can be read directly by computers. Out of this development, a plethora of measures to increase efficiency of utilities or industry can be derived as Lim et al. explain in Chap. 4 references.

Digital Twin: The aggregation of virtual representation of distinct physical entities in cyber-space (Siemens 2018) provided by for instance smart metering infrastructure, the IoT and cloud computing, are commonly referred to as Digital Twin. This approach facilitates many beneficial developments such as predictive maintenance of machinery or process efficiency monitoring. GE (2018) Bishop sets out how Digital Twins can aid the operation and planning of transport networks in Chap. 18, Schmitt and co-workers present experiences for cooling a tropical agglomeration Chap. 17 while Shah and co-workers elicit their experience with optimising complete urban energy systems (Chap. 15).

2.2 Cyber-Physical Systems

These are the main technologies that enable the creation of complete cyber-physical systems, which are ultimately an intelligent orchestration of these break-through technologies. As an example, imagine that advanced metering infrastructure combined with edge computing monitor and control a certain process providing the data—via the cloud—to a digital twin. The digital twin utilises the data received via a Big Data approach to facilitate for instance predictive maintenance. The CPS therefore facilitates the machine-to-machine communication—via smart metering, edge computing and the IoT—and simultaneously provides this data to a digital twin

Fig. 2.1 Architecture of
intelligent cyber-physical
system

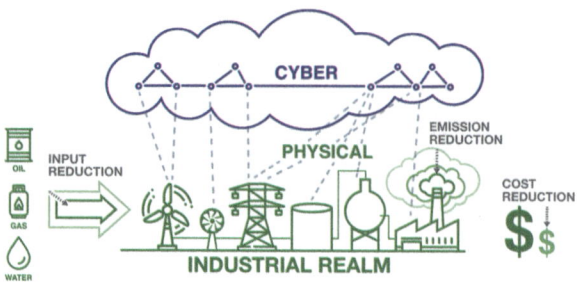

for Big Data analysis in the cloud. From the Big Data analysis system-wide improvements can be deducted, which are fed back to the real system via the cloud and the IoT where they are implemented through edge computing. Figure 2.1 shows a simplified visual scheme of CPS. The synergistic effects that CPS can provide via the orchestration set out above are a topic of the following chapter in which expert give their opinion on the scale of possible improvements. This, however, is not the end of the line as two technologies that will certainly hypercharge the effectiveness of CPS, quantum computing and artificial intelligence, are in the starting blocks.

Quantum Computing: Computers that utilise quantum phenomena such as superposition and/or entanglement to perform computation are referred to as quantum computers. These novel computers will be substantially faster than classical computers, the gap between the two classes of computers is referred to as quantum supremacy. This additional speed is the prerequisite for the next step, artificial intelligence. It will be discussed briefly hereafter and in more detail in the subsequent chapter.

Artificial Intelligence: The ability of computers to perform tasks commonly associated with intelligent (biological) beings is referred to as artificial intelligence (AI). The term is frequently applied to the project of developing systems endowed with the intellectual processes characteristic of humans, such as the ability to reason, discover meaning, generalize, or learn from past experience. In specific tasks, such as playing chess or Go as well as predicting protein folding, AI already beats the human mind, while in general cognitive tasks the biological mind is superior. All this could, however, change rapidly due to advances in AI research as well as the advances in quantum computing; the result of these advances could lead to artificial general intelligence (AGI) and hence to conscious machines. Both AI and AGI are discussed in detail in the subsequent chapter. Is there a consensus that these technologies will facilitate the decarbonisation of the global economic system? The US Department of Energy's Clean Energy Smart Manufacturing Innovation Institute (CESMII) and the European Strategic Energy Technology Plan (SET Plan) put digitalisation and AI at the core of their decarbonisation strategies which substantiates the scientific claim of the importance of digital innovation. As a result, it is urgent to initiate a thorough discussion of how intelligent CPS technologies (e.g. IoT, AMI, ML combined with AI) can be applied in the transition to a net-zero economic system.

The transition to net-zero emissions that was announced by countries such as the UK, Switzerland, Japan among others is a complex, long-term challenge that needs collaborative contributions from low-carbon power provision, energy efficiency enhancement, storage adoption and related areas. Despite the notable progress in these areas, much remains to be achieved to meet emission targets. Intelligent CPS technologies could not only hypercharge such advancements and therefore accelerate the so-called Race to Zero, but also reduce the enormous costs that are anticipated for this transition. Hereafter, use cases on the environmental and economic impact are presented in order to illustrate the potential impact of CPS on decarbonisation potential and cost; from this, the impact of digital technologies on the renowned marginal abatement cost curve (MACC) is estimated.

2.3   Economic and Environmental Impact Assessment

2.3.1   Case Studies

Quintessential to all emission-reduction endeavours is the decarbonisation of electricity supply as electricity is either directly or indirectly embedded in all produce affecting the producer's carbon footprint (Holdway et al. 2010). Moreover, the electricity sector is going through a significant transformation because traditional boundaries between the various branches of energy supply sectors such as heating, cooling, and transport are blurring due to electrification (Inderwildi and King 2012). Therefore, the decarbonisation of electricity provision will positively affect all other sectors and this effect is reinforced owing to electrification endeavours. Due to the essential nature of electricity for overall decarbonisation of economic activity, an economic impact assessment of use cases will be presented hereafter to illustrate the environmental and economic impact of CPS. Firstly, *ad-hoc* calculations of emissions and costs savings for the representative examples outlined in will be presented hereafter. Secondly, representative examples from the literature are examined to quantify the benefits of CPS.

Illustrative calculations:

1. **Building management system**: EIA estimates that building energy use will be responsible for 30% of global energy use by 2050, which corresponds to 1000 TWh (Energy Information Administration 2017). According to the analysis of California's Independent System Operator (CAISO), 10% energy savings can be achieved through implementing BMS-related energy management and demand side response techniques, enabling the potential of 100 TWh reduction in electricity use (International Energy Agency 2017). Reducing the need for this portion of electricity results in a potential abatement of 22 Mt of CO_2 (eq.) as well as a reduction of costs of approximately 16 bn US$.
2. **EV charging**: The IEA's Global EV Market Outlook anticipates over 120 million EVs on road in 2030, resulting in an overall energy demand for EV charging that

accounts for 6% of global power demand, (approximately 200 TWh) (Cazzola et al. 2016). The analysis shows that coordinated EV charging can reduce the electricity demand by 40% (Qian et al. 2011), resulting in 80 TWh total savings. Based on current electricity prices and emission intensities, the benefits of CPS in EV charging can be quantified as 18 Mt CO_2 emission abatement and 13 bn US$ saving respectively.

3. **Renewable forecasting**: Based on a projection of the EIA, energy provision by intermittent renewables (e.g. wind and solar) has the potential to reach 11,500 TWh by 2050 (Energy Information Administration 2017). Again, according to a case study by California's Independent System Operator (CAISO), a 15% increase in renewable penetration can be expected with a 10% forecast performance improvement (Hodge et al. 2015), which equals to 1725 TWh electricity generation increase. The corresponding CO_2 emission reduction potential and financial optimization can reach 380 Mt and 276 bn US$ respectively.

4. **Power system optimization**: For the proposed integrated energy management tool—J-Park Simulator—previous studies have shown that by optimizing the power and heat cogeneration system on Jurong Island Singapore, the annual power generation can be reduced from 19 to 12 TWh providing a staggering reduction of 63% (Rigo-Mariani et al. 2019). Using Singapore's power emission intensity and electricity price (i.e. 431 kg/MWh and 0.12US$/kWh) as a benchmark (Kannan et al. 2007), it is estimated that 3 Mt of CO_2 emissions could be achieved while 840 m US$ could be saved.

These illustrative sample calculations based on the representative cases presented herein already show the substantial economic gains and environmental benefits stemming from the application of CPS. In Part III of this book, experts from various fields will discuss the emissions and cost savings potential of CPS in more detail. One negative impact of CPS is their inherent energy consumption and hereafter the trade-off between benefits and costs will be discussed.

Trade-off between energetic costs and benefits: The energy savings potential of CPS obviously goes hand-in-hand with energy utilised by the system and net energy savings have to be assessed to gauge the overall impact. Thiede studied a cyber-physical production system called EnyFlow that monitors energy demand data. It collects this data with a resolution of 1 second for machinery (10 W, 220 days per year) while utilising a desktop computer (150 W, 8760 hours per year), and tablet as well as several sensors. Introducing EnyFlow increases the energy invested by 7.7% while reducing energy needs by 20% leading to a net improvement of 12.6%. The economic break-even, i.e. the return on investment, for EnyFlow is achieved within the first year according to Thiede (2018). The supporting feasibility diagram is depicted in Fig. 2.2 More details will be given in Chap. 8 by Thiede.

As decision support for assessing CPS, feasibility diagrams are essential. Based on the EnyFlow case, Fig. 2.2 shows favorable and non-favorable areas for CPS based on the absolute potential (in kWh) that is addressed combined with the necessary relative improvement impact over a defined time frame. The isopleths mark the break-

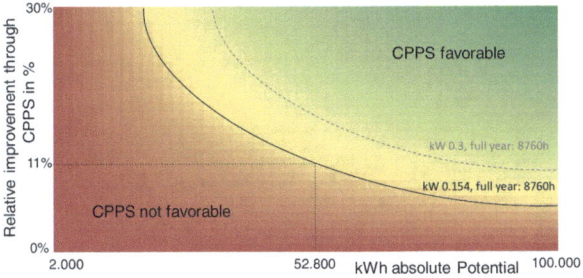

Fig. 2.2 hICPPS environmental feasibility diagram based on EnyFlow Thiede (2018)

even line. With that, for a given production situation (with its potential) necessary relative improvements to achieve a break-even in a given time frame can be derived. This example clearly illustrates that CPS can provide economic and environmental benefits.

2.3.2   Impact on the Marginal Abatement Cost Curve

In 2007, long before the advent of CPS, McKinsey and Company published the marginal abatement cost curve (MACC), a curve that illustrates both the marginal cost of abatement as well as the abatement potential of certain technologies. The emergence of CPS technologies has brought further improvements that will have to be considered in an evolution of the MACC in order to further establish it as a discussion tool for climate mitigation strategies; hereafter, it will be argued that the MACC was further altered by CPS and has to be updated continuously due to the rapid improvements driven by digital technologies.

CPS have altered both the abatement potential and the economics of selected decarbonization technologies, because CPS improve efficiency, reduce risks and optimise overall processes. Figure 2.3 shows a simplified version of the original MACC with each technology characterized by the abatement potential and cost (top), the impact of CPS on the MACC resulting in a 20% increase in abatement potential (bottom) (Vogt-Schilb and Hallegatte 2014). In the baseline scenario, the marginal abatement potential and cost of selected sectors by 2030 (including building sector, petrochemical, iron and steel sector, solar, wind, coal CCS, BECCS and hybrid/electric vehicle as shown in the legend in Fig. 2.3) is taken from the original publication, the aggregated CO_2 mitigation potential from these sectors is estimated at 7.7 Gt (The Global Greenhouse Gas Abatement Cost Curve 2009; Kesicki and Ekins 2012; McKitrick 1999).

In the CPS scenario, the aforementioned CPS technologies are assumed to be applied, leading to significant increases in abatement potential as well as reductions in cost as noted by striped areas in the middle of Fig. 2.3. A few illustrative exam-

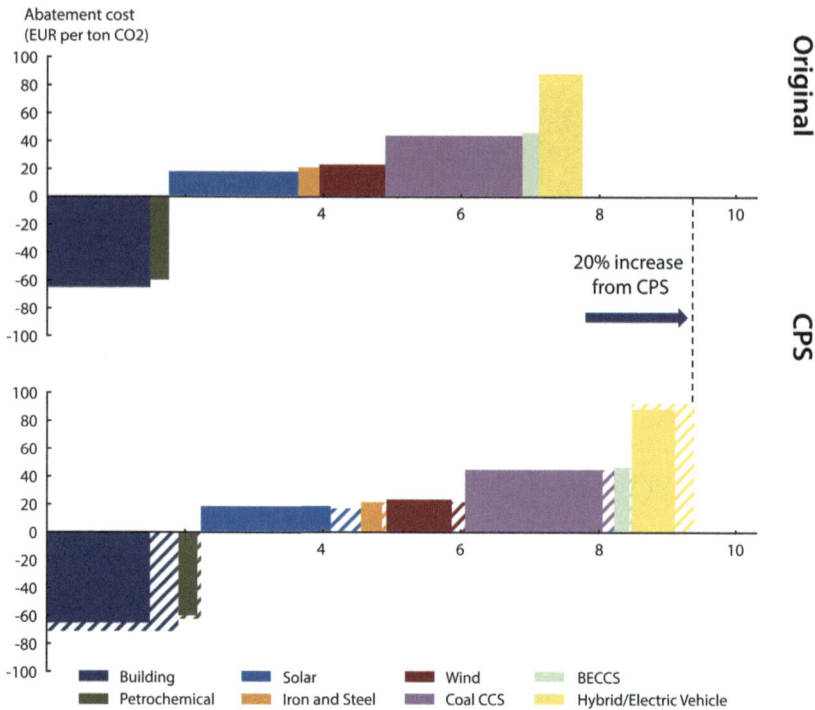

Fig. 2.3 Impact of CPS technologies on the marginal abatement cost of selected decarbonization technologies in energy transition. Shown in the figure is the marginal abatement cost of selected decarbonization technologies (represented by different colors) without CPS technologies (top The Global Greenhouse Gas Abatement Cost Curve 2009), with CPS technologies (middle Electric Vehicle Initiative 2018; Hodge et al. 2015; Strategic Energy Technologies Information System 2018), and with intelligent CPS technologies (below Vogt-Schilb and Hallegatte 2014; Kesicki and Ekins 2012; McKitrick 1999)

ples clarify the basis of the newly rendered MACC: Buildings are forecasted to be responsible for 20% of global energy use and building energy management is likely to reduce 10% through CPS-optimisation resulting in a 3% reduction in global energy use (International Energy Agency 2017). EV charging will be responsible for 6% of global electricity demand and coordinated charging could reduce demand by 40% consequently reducing electricity demand globally by 2.4% (Qian et al. 2011). CPS-based integrated energy management can reduce energy use of industrial complexes by 63% and thereby increase CO_2 emission abatement potential by this factor (Rigo-Mariani et al. 2019). A full overview over the data utilised to establish new MACC curve (bottom of Fig. 2.3) including references to the original literature can be found in Inderwildi (2020).

From the survey outlined in the preceding sections and the illustrative examples above, it can be concluded that state-of-the-art CPS technologies will continue to be integrated in different sectors in the coming years and the delivery of economic and environmental benefits is highly likely. Moreover, future developments of in CPS technologies, such as the combination of AI with CPS, will be discussed by experts in the remainder of this book. Based on the analysis presented within this chapter, it is estimated that by integrating both current CPS technologies and future CPS technologies into the investigated sectors, the CO_2 abatement potential could be increased from 7.7 to 9.4 Gt (20% increase). In particular, the largest increase comes from hybrid/electric vehicles, the building sector as well as solar and wind generation, for which the CO_2 abatement potential could increase by 40%, 28% and 20% respectively.

While the estimates used for the establishment of the revised MACCs displayed in Fig. 2.3) are subject to distinct uncertainties, the economic assessment of CPS technologies combined with the abatement potential speaks for itself: CPS can catapult us on unforeseeable decarbonisation pathways while saving money. In order to meet emission targets and restrict the increase of global average temperatures to 1.5 °C, CPS are indispensable. In the subsequent chapter, artificial intelligence will be discussed and the chapters in Part III will give experts the opportunity to give their assessment of the impact of intelligent CPS on both economic improvements as well as their climate change mitigation potential. Part IV will be concerned with the risks posed by the proliferation of AI and again will give experts, scholars and practitioners the opportunity to put forward possible means to mitigate these risks.

2.4 Conclusions

In this chapter, the cyber-physical systems (CPS) as well as their subsystems were discretised. CPS are defined as high-level orchestrations of various cutting-edge digital technologies creating a digital representation of the physical world. This digital representation enables the enhanced operation and control of the physical system using advanced optimisation tools as well as novel digital technologies. Among these technologies are big data, machine learning, IoT, AMI, blockchain and the semantic web. This orchestration will lead to synergetic benefits in terms of economic improvements and environmental sustainability that go way beyond the potential of isolated digital technologies.

Case studies were used to illustrate the benefits of CPS and its sub-systems for the economic and environmental viability of various production systems. In the case of renewables, CPS will promote the integration of the intermittent energy source in existing energy systems, e.g. through computer-aided renewable resources identification and renewable energy forecasts improved by machine learning technology. In the case of energy efficiency, CPS are having an impact that is highly likely to increase significantly over the coming years; this effect is exemplified using advanced building management systems, the optimisation of data centres and enhanced demand side management. Last but not least, CPS will be critical for facilitating energy storage especially when considering cross-sector power-to-X. This effect is illustrated using CPS-enhanced smart charging of electric vehicles and blockchain-enabled P2P energy trading.

These case studies illustrate that CPS will not only be beneficial for economic optimisation, but will deliver environmental advantages like emission abatement. *Ad hoc* calculations of the monetary benefits and emission savings potential illustrate and prove this point. CPS are already having an impact on CO_2 emissions and the economic viability of industry; it was shown that even in sectors that are used to improvements in the single-digit percentages, CPS can enable improvements of 30% and beyond. A clear illustration of the cumulative benefits of intelligent CPS is given in the revamped MACC curve—abatement potential can be improved by 56% while costs can be reduced by more than 30%. Our study therefore also provides an overdue update to the MACC and further establishes it as a tool for practitioners and policy makers concerned with decarbonisation efforts.

Our assessment also clearly shows that supporting CPS is also an obvious climate change mitigation strategy. Governments should support CPS that drive energy efficiency as part of their decarbonisation strategies; from a cost-benefit perspective CPS deliver emission savings at a cost far below that of traditional mitigation strategies. Policy makers should consider incentives for CPS that optimise not only energy provision, but also industrial production and transport.

All these benefits have clear downsides as CPS are connecting cyber-threats with the real, physical world and therefore cyber-crime could have a more direct impact. Policy makers have to continuously review CPS and in order to foster benefits and safeguard risks by simultaneously supporting cyber-security measures while regulating critical areas. Policies for this have to be adapted continuously in a cycle much shorter than the current one. In Part IV, experts will elaborate on the regulatory and governance issues of CPS.

2.5 Outlook

It was shown that CPS are already having an impact on emissions while there are indeed obstacles with regards to implementation and therefore, the beneficial influences of CPS are not fully harnessed yet. The impact of CPS will likely increase in the short-term and certainly will affect energy provision and industrial production in the long-term. The actual impact will depend on the real-world adoption of novel technologies, such as electric vehicles, renewables and distributed storage, but in any case will be significant. Ekholm and Rockström estimate that in energy provision, manufacturing, agriculture and land use, buildings, services, transportation and traffic management combined, an emission reduction of 15%, i.e. one-third of the 50% reduction required, could be delivered by digitalisation alone (Ekholm and Rockström 2019). In a related assessment, the energetic costs and benefits of digitalisation are examined and it is concluded that the benefits clearly outweigh the costs as the digital sector itself reduces its carbon footprint continuously (TWI2050 Initiative 2018).

Acknowledgements This research is funded by the National Research Foundation (NRF), Prime Minister's Office, Singapore under its Campus for Research Excellence and Technological Enterprise (CREATE) programme. We would also like to acknowledge Ms. Louise Renwick and Dr. Andrew Breeson for helping us improve our English writing.

References

Berners-Lee T, Hendler J, Lassila O et al (2001) The semantic web. Sci Am 284(5):28–37

Cazzola P, Gorner M, Schuitmaker R, Maroney E (2016) Global electric vehicle outlook 2016. Technical report, International Energy Agency

Ekholm B, Rockström J (2019) Digital technology can cut global emissions by 15%. Here's how. https://www.weforum.org/agenda/2019/01/why-digitalization-is-the-key-to-exponential-climate-action/. Last accessed 15 November 2019

Electric Vehicle Initiative (2018) Global EV outlook 2018: towards cross-model electrification. Technical report, International Energy Agency

Energy Information Administration (2017) International energy outlook 2017: reference case projections for electricity capacity and generation by fuel. Technical report, U.S

GE (2018) GE Predix. https://www.predix.io/. Last accessed 15 May 2019

Gubbi J, Buyya R, Marusic S, Palaniswami M (2013) Internet of things (IoT): a vision, architectural elements, and future directions. Futur Gener Comput Syst 29(7):1645–1660. https://doi.org/10.1016/j.future.2013.01.010

Hodge BM, Florita A, Sharp J, Margulis M, Mcreavy D (2015) Value of improved short-term wind power forecasting. Technical report, National Renewable Energy Lab (NREL), Golden, CO (United States)

Holdway AR, Williams AR, Inderwildi OR, King DA (2010) Indirect emissions from electric vehicles: emissions from electricity generation. Energy Environ Sci 3(12):1825–1832. https://doi.org/10.1039/C0EE00031K

Inderwildi O, King SD (2012) Energy, transport & the environment: addressing the sustainable mobility paradigm. Springer

Inderwildi OR (2020) The impact of intelligent cyber-physical systems on the decarbonization of energy. Energy Environ Sci 13:744–771

International Energy Agency (2017) Digitalization and energy. Technical report, IEA

Kannan R, Leong K, Osman R, Ho H (2007) Life cycle energy, emissions and cost inventory of power generation technologies in Singapore. Renew Sustain Energy Rev 11(4):702–715. https://doi.org/10.1016/j.rser.2005.05.004

Kesicki F, Ekins P (2012) Marginal abatement cost curves: a call for caution. Clim Policy 12(2):219–236. https://doi.org/10.1080/14693062.2011.582347

McKitrick R (1999) A derivation of the marginal abatement cost curve. J Environ Econ Manag 37(3):306–314. https://doi.org/10.1006/jeem.1999.1065

National Grid System Operator UK (2018) Future Energy Scenarios (FES). http://fes.nationalgrid.com/fes-document/. Last accessed 15 May 2019

Qian K, Zhou C, Allan M, Yuan Y (2011) Modeling of load demand due to EV battery charging in distribution systems. IEEE Trans Power Syst 26(2):802–810. https://doi.org/10.1109/TPWRS.2010.2057456

Rigo-Mariani R, Zhang C, Romagnoli A, Kraft M, Ling KV, Maciejowski JM (2019) A combined cycle gas turbine model for heat and power dispatch subject to grid constraints. IEEE Trans Sustain Energy 11(1):448–456. https://doi.org/10.1109/TSTE.2019.2894793

Samuel SSI (2016) A review of connectivity challenges in IoT-smart home. In: 2016 3rd MEC international conference on big data and smart city (ICBDSC), pp 1–4. https://doi.org/10.1109/ICBDSC.2016.7460395

Siemens (2018) Siemens MindSphere. https://www.siemens.com/global/en/home/products/software/mindsphere.html. Last accessed 15 May 2019

Strategic Energy Technologies Information System (2018) Digitalisation of the energy sector. Technical report, The Europe Union

The Global Greenhouse Gas Abatement Cost Curve (2009) Pathway to a low-carbon economy. Technical report, Mckinsey & Company

Thiede S (2018) Environmental sustainability of cyber physical production systems. Procedia CIRP 69:644–649

Törngren M, Bensalem S, Cengarle M, McDermid J, Passerone R, Sangiovanni-Vincentelli A (2014) Cyber-physical European roadmap and strategy. Cyber-Physical European Roadmap and Strategy D5. 1, Technical report

TWI2050 Initiative (2018) Transformations to achieve the sustainable development goals. Technical report, International Institute for Applied Systems Analysis

Vogt-Schilb A, Hallegatte S (2014) Marginal abatement cost curves and the optimal timing of mitigation measures. Energy Policy 66:645–653. https://doi.org/10.1016/j.enpol.2013.11.045

Chapter 3
Artificial Intelligence

Oliver Inderwildi and Markus Kraft

Abstract Digital technologies are already impacting every area of life, from economic and commercial activity to social interactions and political processes. These significant impacts are likely to be outshone when the next technological quantum leap appears on stage: artificial intelligence (AI). In specific tasks, most prominently playing games such as Chess or Go, artificial intelligence already eclipses human ability. Advancements in deep learning, however, are broadening the skillset of AI and the ramifications for us are unforeseeable. To elicit the potential impact, the general popular literature on AI and its capabilities is summarized in this chapter. The evolution of machine intelligence from the specific AI 1.0 to the general AI 3.0 is discussed while for the latter, the potential perils and downsides are balanced with its enormous potential for good.

3.1 Introduction

The advent of artificial intelligence (AI) has sparked a ferocious debate on its risks and benefits. While it is clearly important to be aware of the risks, the potential benefits have to be considered as well. The British polymath James Lovelock, for instance, claims that humanity is leaving the Anthropocene and is entering the Novacene, an age in which the technology created by humans moves beyond our control and generates intelligences greater than ours. Humans are the chosen people, chosen in a Darwinian selection process for our intelligence, which in turn enabled us to create our own Darwinian successor, artificial intelligence (AI), according to the centenary. AI will help us to solve humanity's and the planet's problems, but humans will lose their status as the most intelligent species (Lovelock 2019).

In *Homo Deus* the Israeli thinker Yuval Noah Harari refers to this as the "The Great Decoupling" i.e. the decoupling of intelligence and consciousness (Harari 2015). Nick Bostrom draws a darker picture and assumes that as soon as AI has

O. Inderwildi (✉) · M. Kraft
Cambridge CARES, Singapore, Singapore
e-mail: oliver.inderwildi@scnat.ch

© Springer Nature Switzerland AG 2022
O. Inderwildi and M. Kraft (eds.), *Intelligent Decarbonisation*, Lecture Notes in Energy 86, https://doi.org/10.1007/978-3-030-86215-2_3

surpassed the human intellect, it may identify humanity as superfluous and thus exterminate it (Bostrom 2014).

MIT's Max Tegmark has a more human-centered point of view and claims the first ultra-intelligent machine is the final invention humans have to make. He defines three levels of life (Tegmark 2017):

- Life 1.0: it can replicate, e.g. cells
- Life 2.0: it can design its own software, e.g. a human learning a language
- Life 3.0: it can design its own hardware and software

We are approximately at Life 2.1 as we can alter our hardware to a certain degree by replacing teeth, curing diseases and replacing joints or limbs of the human body. By designing our own hardware, i.e., by curing diseases and replacing limbs, we are already beating evolution! Meanwhile, the replacements are often superior to the original designed by millennia of evolution. This is illustrated by the opposition of leading sprinters during the 2012 London Olympics to their challenger Oscar Pistorius; the sprinters claimed that the double-amputee, who uses carbon-fibre prosthetics as lower legs, uses less energy in a sprint and hence has an unfair advantage over them. Hence, for the *specific* task of sprinting, the human-developed version beats the version already designed by evolution. The same holds for AI in specific tasks: AI purpose-built for tasks such as playing Chess, Go or Jeopardy does not leave humans a chance as illustrated by the 1997 defeat of Chess Grandmaster Garry Kasparov against IBM's DeepBlue, the 2008 defeat of all-star Jeopardy players by IBM's Watson and the 2016 dismantling of professional Go player Lee Sedol by AlphaGo (Kurzweil 2012). Also, in other *specific* areas such as the interpretation of medical images (Ranschaert et al. 2019) and the secure steering of an automobile, AI already trumps the human mind on many occasions. However, a surgeon can probably drive a car and play Chess, Go and Jeopardy; AI still lacks this flexibility but closes in rapidly. Why have advances in AI taken off over the last few years and what is next to come in this fascinating development? All this is part of the subsequent sections.

3.2   Main Text

3.2.1   The Advent of AI

Since the 1980s, the AI revolution was always deemed to be five years away, but now developments are really picking up speed, mainly pushed by tech giants in Silicon Valley and Beijing's Avenue of Entrepreneurs (Lee 2018). What caused this significant increase in speed of AI development? Two main developments are, in our opinion, key: firstly, deep learning based on neural networks and secondly, the rapid increases in computational power and availability of large amounts of data.

In the early days of AI research, two schools of thought existed: the rule-based approach and the neural network approach. The rule-based approach tried to teach machines to think by encoding them with a set of logical rules, which works well for well-defined problems—think of a chess computer (not DeepBlue!). However, when problems become too complex, this approach does not suffice as too many logical options would need to be encoded—an enormous and inefficient decision tree would be generated. This work—to some extent—is still ongoing and is referred to as AI 1.0. More recently, the purely data-driven AI research, mainly using neural networks, has been inspired by the workings of the human brain. This approach mimics the architecture of the human brain by creating layers of artificial neurons that process information. These neural networks are then repeatedly fed with information—pictures, sounds, games etc.—and left to spot patterns with no human interaction, so-called deep learning. What neural networks require to learn and develop has only recently become available in sufficient amounts, which brings us to the second key development: computational power and data availability.

Advances not only in computational power but also in its price are hard to fathom. In 1965 Gordon Moore anticipated that the number of transistors on a microchip doubles approximately every year. The so-called Moore's Law holds since 1975 (blue line, Fig. 3.1) and, simultaneous to the increase in the power of a single microchip, the computational cost calculated in US$ per gigaflop has been reduced to mere cents (green line, Fig. 3.1). *Nota bene*: both lines are on logarithmic scales. Hence, in the early days of AI research, neural networks were doomed to fail as costs were too high while power was not sufficient.

Moreover, the availability of data and storage is not a hindrance anymore: the internet creates more than an exabyte of data every day, that is 10^{18} bytes; the overall

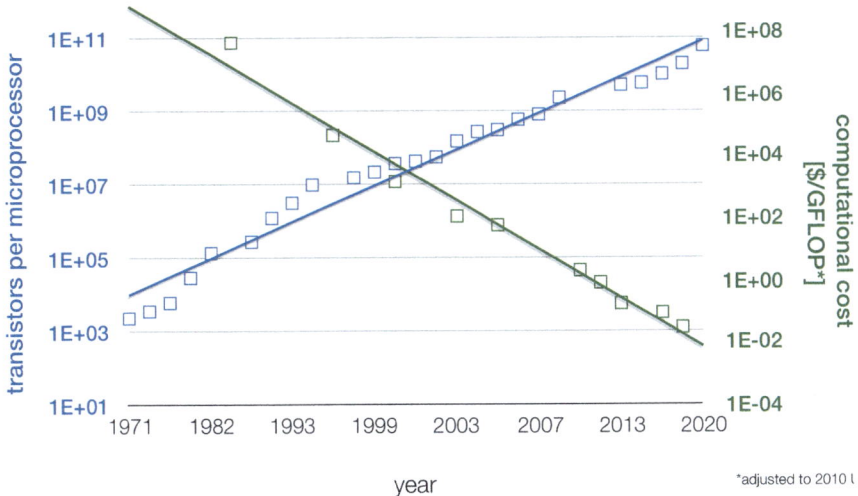

Fig. 3.1 Moore's Law on the evolution of microprocessor power (blue) and the reduction of computational cost (green); Data via Our World in Data

data generation is estimated to be 2.5 exabytes. For comparison, the age of the world is approximately 0.43 exaseconds. The same is true for storage; a regular mobile phone today holds millions of times more data than the computers NASA used for the moon landing. Moreover, we freely share data on the internet and social media so tech giants have copious amounts of data with which to train their neural networks.

However, one of the key criticisms countering such a purely data-driven approach is explainability; in other words, it is not known "why" a certain outcome occurs. This current, highly successful phase of AI research has been termed AI 2.0. Many AI researchers have taken on the challenge and are now developing algorithms and methods that will eventually lead to explainable AI. It is envisaged that these methods will combine human thinking with elements of both AI 1.0 and 2.0 and will result in very powerful technologies—when this happens, AI 3.0 will have arrived. AI 3.0 will represent a collective intelligence formed not by a group of people, but by humans interacting with machine intelligence. This approach would make AI explainable and accessible while mitigating the risks posed by general AI (Malone 2018).

3.2.2 Getting Deep

Training neural networks was a big breakthrough, however, to progressively extract higher-level features from raw data, many layers of the neural network are required. Also, the more complex the problem-to-be-solved gets, the more layers the neural network requires. More layers add complexity to the neural network, which creates inefficiencies and might augment error propagation that limits its effectiveness.

In a stroke of genius, a backpropagation algorithm was added to the communication between so-called hidden layers, which made communication between layers more efficient due to a reduced number of errors—deep learning was born (Bengio et al. 2015). These hypercharged deep learning networks outperformed other approaches at many tasks and neural networks went from fringe to state-of-the-art. This was in 2012 and when you speak to Alexa or Siri today or sit in a self-driving Tesla or use the translator Deepl, you make use of backpropagation and deep learning. Deep learning was also used by DeepMind's AlphaGo to beat Lee Sedol.

Deep learning has also been successfully used by DeepMind to improve energy efficiency. Cooling data centres is a highly energy-intense process and consequently, a significant financial and potentially environmental burden. Cooling data centres is complex due to the manifold interactions between different types of equipment, the building and the surroundings (including changing weather conditions). Owing to this complexity, traditional, rule-based approaches do not work perfectly. A deep learning algorithm for adaptive optimisation of data centres was devised and could reduce energy demand from cooling by 30% (green line) using more than 80 million training examples (blue line, Fig. 3.2). Given that Google's data centres already are highly optimised, this is clearly an enormous achievement. The approach is now delivering cost, energy and emissions savings worldwide. The efficiency gains outlined above have economic and environmental ramifications which will be addressed throughout

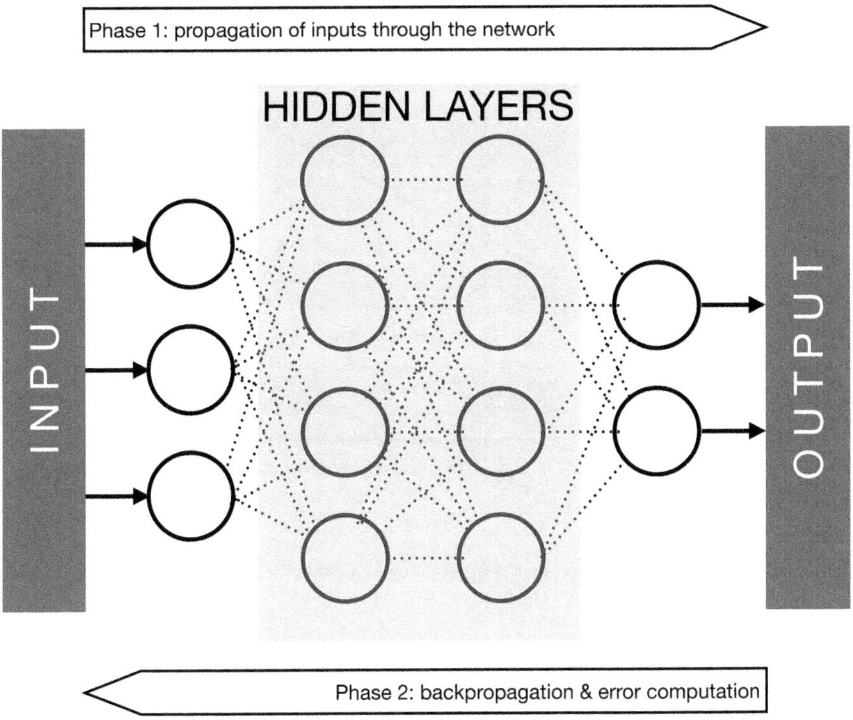

Fig. 3.2 A neural network with back propagation. In Phase 1, inputs propagate through the network (left to right); in Phase 2, a backpropagation checks for errors

this book. To start, however, a short back-of-the-envelope calculation: the energy consumption of global data centres is forecasted to grow to 3000 TWh by 2025 (International Energy Agency 2017). Since DeepMind's technology has the proven potential to reduce energy consumption of servers and data centres by 30%, 900 TWh of electricity generation could be saved. Using today's electricity price and carbon footprint of California (215 kg/MWh and 0.16 US$/kWh resp.) where most servers are located, these efficiency improvements would equate to a staggering emission reduction of 193 Mt as well as 144 bn US$ cost savings (Inderwildi et al. 2020). Such examples clearly illustrate the environmental and economic benefits the AI area can deliver (Fig. 3.3).

3.2.3 Leaving Specificity

Google's Ray Kurzweil forecasts that machines will reach a human level of intelligence before the end of this decade while technological singularity, the hypothetical point when technological progress can no longer be controlled by humans,

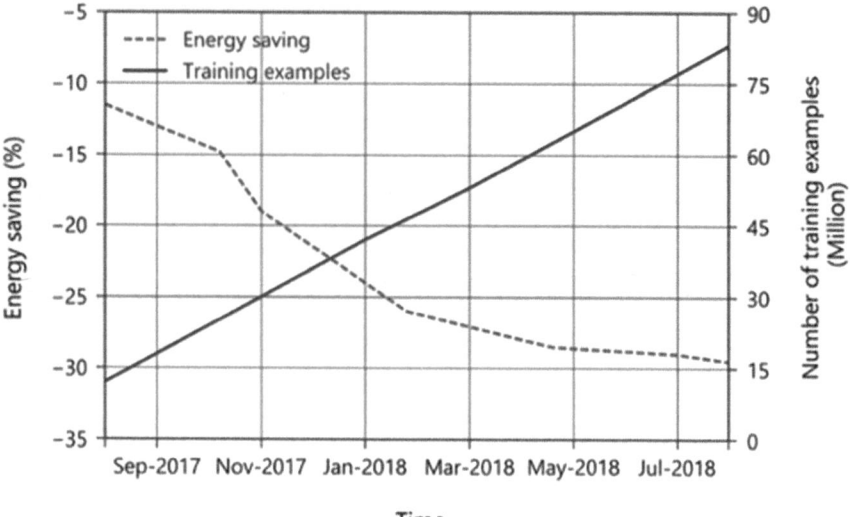

Fig. 3.3 Energy savings achieved by a deep learning algorithm for cooling Google's data centres

will be reached before 2045 (Kurzweil 2012). Singularity is a critical inflection point at which AI as it is known today, vide supra, will transition to artificial general intelligence (AGI). AGI allows a machine to apply knowledge and skills in different contexts; such a machine intelligence closely mirrors human intelligence by providing opportunities for autonomous learning and problem solving. Will it think like a human being? Richard Feynman famously predicted that AGI would come, but that its thoughts and potential consciousness will differ from ours (Leighton and Feynman 1985).

Kurzweil deducts from a set of thought experiments that the human brain draws its power from a hierarchical set of pattern recognition circuits in the neocortex—300 million to be precise! He uses these deductions as the basis of his Pattern Recognition Theory of Mind and suggest that computational models of these biological algorithms could create human-like artificial intelligence (Kurzweil 2012). Deep learning clearly provides the foundation of this next step; however, even more computational power will be required, computational power that cannot be provided by conventional computers. A disruptive technology that will be able to overcome this hurdle can already be seen on the horizon: quantum computing.

This novel computational approach utilises quantum phenomena such as superposition of matter and quantum entanglement. This method enables the quantum computer to move away from the byte (0 or 1) and introduce the fluid quantum bit, a short qubit that can be both 0 or 1 in different proportions (0–1). The fluidity of the qubit provides the quantum computer with not only almost unlimited storage, but an enormous acceleration of computational steps (Hidary 2019). Google announced that its pilot quantum computer is 100 million times faster than a conventional supercomputer. A critical development step here will be the proof of quantum supremacy; the proof that a quantum computer can solve a problem that a conventional computer cannot solve. Proof of quantum supremacy would e.g. enable the simulation of complex quantum systems such as biological molecules, or offer a way to factor incredibly large numbers, thereby breaking long-standing forms of encryption. Further improvements in neural networks, deep learning and quantum computing could indeed prove Kurzweil right in that technological singularity is nigh.

The authors, however, agree with Max Tegmark that we must answer crucial questions before we develop AGI as the risks might outweigh the benefits, vide infra (Tegmark 2017).

3.2.4   The Economics of the AI Age

Advances in AI will affect productivity, growth, inequality, market power, innovation and employment; sometimes for the good, sometimes for the bad (Agrawal et al. 2019). In *The Road to Wigan Pier* George Orwell foresaw that the increasing automation of the industrial age will render the commoner less valuable to society as many physical tasks will be completed by machines—capital had won over labour, Orwell concluded (Orwell 1937). Harari argues that liberals uphold the free market and democratic elections due to a belief that every human is uniquely valuable and the ultimate authority lies with the human collective. In war and on production lines, every human counts, however, in cyberwars and on automated production lines the value of a human is not as unique. As soon as creative and intellectual tasks are transferred to intelligent machines, human uniqueness and value is called into question. For the functioning of the economy, intelligence is necessary while consciousness is optional. This decreased utility of the individual jeopardises the alliance of capitalism and liberalism, according to Harari, and could consequently question the value of the individual (Harari 2015).

It is clear that AI will render many jobs obsolete just as the automation during industrialisation did: chatbots replaced many call centre workers, robo-advisers outperform investment managers, translation software clearly impacts the business of translators. A business consultancy estimated in 2019 that AI might add 7.4% or 15.7 trillion US$ to GDP forecasts (PwC 2019) (see Fig. 3.4). However, due to the Covid-19 pandemic, significantly faster rates of digitalisation have been observed and the authors believe that this number will have to be corrected upwards.

The world might end up in an Athenian scenario in which everyone goes about their leisure while robots and AI run the economy. In such a scenario, it must be ensured that humanity has a tight grip on the rudder and does not degenerate due to lack of mental exercise. This is a very complicated scenario as we will see in the subsequent section.

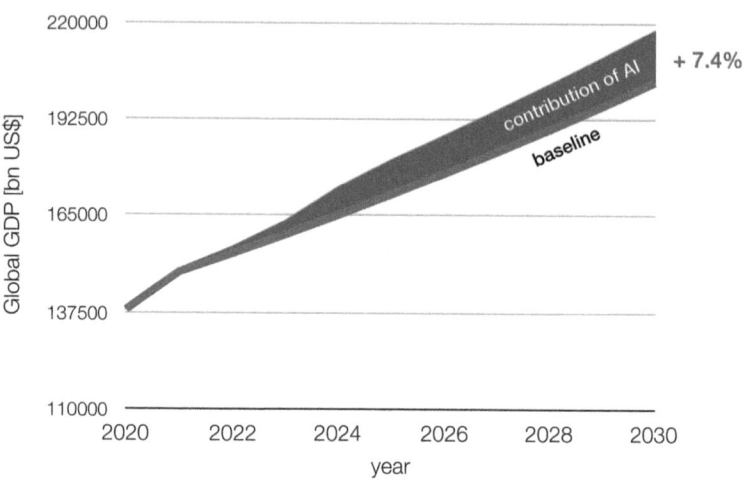

Fig. 3.4 A 2019 forecast for additional economic growth through AI; *Source* Microsoft and PwC

3.2.5 The Risks of AI

As mentioned in the introduction, with great opportunity comes great risk. Nick Bostrom, in his book *Superintelligence: Paths, Dangers, Strategies*, addresses the so-called "AI control problem" (Bostrom 2014). Once human-like AGI is developed, the singularity (or superintelligence) is not far off and such a scenario would be—by definition—almost impossible to control. A superintelligence would proactively restrain all attempts to be turned off, would certainly set sub-goals or indeed alter the goals set for itself. Control of this would be impossible—the AI control problem. One could attempt to make sure that the superintelligence's goals are aligned with humanity's in order to ensure that it does not alter the overarching goal, but this is also difficult—if not impossible—to achieve due to unforeseen and undesirable consequences of goals. The question of AI control has to be answered before developing—or allowing the development—of AGI as this poses an existential risk for humanity. While an existing superintelligence cannot be controlled by humans, its development can be prevented. Humans must make sure that the AI control questions are answered before the creation of a superintelligence is allowed to take place.

As laid out in the introduction, we might create technology that puts an end to humanity's problems or indeed humanity itself. It has to be ensured that AGI's goals are aligned with humanity's goals as AGI lacks a moral compass (Sautoy 2020) and is therefore likely to ruthlessly pursue its *own* goals. We should be proactive rather than reactive and we have to invest drastically in AI safety research given the speed of developments in different areas. The interested reader is referred to the Future of Life Initiative as well as the interview with David Rolnik (Chap. 5).

3.3 Conclusions

Artificial intelligence has entered our everyday lives—and it is here to stay! At specific tasks, such as games like chess, AI beats humans already by a large margin; this specific, non-transferrable intelligence is referred to as AI 1.0. The renaissance of neural networks, and more specifically deep learning, aided by advances in computational power and data availability has provided us with AI 2.0. This machine intelligence learns from massive data sets and deducts patterns, for instance, for voice or facial recognition; the issue with AI 2.0 is the absence of explainability, i.e. the reasons for AI decisions are not known. The creation of explainable AI is at the heart of current research. In particular, the formation of collective machine–human intelligence in the form of knowledge graphs is of interest as this would help to make AI decisions transparent. As soon as collective intelligence is established, AI 3.0 will have been born. This AI transparency will be critical for reasons of safety, fairness and equality when more and more AI decisions will affect everyday life.

Quantum computing and hence the drastic acceleration of computing will almost certainly broaden the capabilities of AI in the long run, i.e. AI capabilities will not be restricted to specific tasks, but will be broadly applicable to challenges ranging from fundamental science to everyday life. Quantum computing will be the ultimate enabler of artificial general intelligence and the singularity.

Once AGI enters this world, a fundamental restructuring of the economic and social system will occur as this type of machine intelligence will reduce the need for humans in the economic systems. This reduction will also have far-reaching consequences for our social systems. Most risk, however, stems from the so-called AI control problem, the issue regarding the containment of a superintelligence. It is argued that the creation of AGI should not be permitted before the AI control problem is solved as this poses an existential risk to humanity. Therefore, proactivity is required and investments in AI safety research are necessary.

3.4 Outlook

AI already has an impact on our daily lives and this impact will continue to grow in all sectors of the economy and our social interactions. In this book, especially the environmental but also the economic input on industry (Chaps. 8, 9, 10 and 12), agriculture (Chapter 26), electricity (Chapter 12), water (Chapter 13), transport (Chapter 18) and cities (Chapters 15 & 17) will be addressed and estimates for improvement potential will be put forward. We will see that AI will not only save emissions and costs now—through operational expenditure reduction—but also in the future through optimisation of capital expenditure.

References

Agrawal A, Gans J, Goldfarb A (2019) The economics of artificial intelligence: an agenda. The University of Chicago Press, Chicago, USA

Bengio Y, LeCun Y, Hinton G (2015) Deep learning. Nature 521(7553):436–444

Bostrom N (2014) Superintelligence: paths. Strategies, Oxford University Press, Oxford, UK, Dangers

Du Sautoy M (2020) The creativity code: how AI is learning to write, paint and think. Fourth Estate Publishing, London, UK

Harari JN (2015) Homo Deus—a brief histroy of tomorrow. Harvill Secker London, UK

Hidary JD (2019) Quantum computing: an applied approach. Springer, Baar, Switzerland

https://futureoflife.org/2017/08/29/friendly-ai-aligning-goals/. Accessed 29 Mar 2021

Inderwildi OR, Zhang C, Wang X, Kraft M (2020) Energy Environ Sci 13(3):744–771

International Energy Agency 2017International Energy Agency (2017) Digitalisation and energy, Paris, France

Kurzweil R (2012) How to create a mind. Viking Penguin, New York, USA

Lee KF (2018) AI Superpowers. Houghton Mifflin Harcourt, Boston, USA

Leighton R, Feynman R (1985) Surely, you're joking, Mr. Feynman. WW Norton, New York, USA

Lovelock J (2019) Novacence. Penguin, Random House, London, UK

Malone TW (2018) Superminds the surprising power of people and computers thinking together. Little, Brown Spark, NewYork, USA

Orwell G (1937) The road to Wigan Pier. Victor Gollancz Ltd, London, UK

PwC H (2019) AI can enable a sustainable future. UK, London

Ranschaert ER, Morozov S, Algra PR (2019) Artificial intelligence in medical imaging. Springer Nature Switzerland

Tegmark M (2017) Life 3.0. Knopf, New York City, USA

Chapter 4
The World Avatar—A World Model for Facilitating Interoperability

Mei Qi Lim, Xiaonan Wang, Oliver Inderwildi, and Markus Kraft

Digitalisation enhances communication and therefore offers new ways to achieve efficiency gains in science, technology and society at large. However, there are still many open questions around how digitalisation can contribute to a more sustainable environment and lifestyle. We believe that knowledge graph technology is a promising candidate with which this can be achieved. In this chapter, we present the World Avatar, a dynamic knowledge graph (dKG), and explain the main underlying concepts and principles. Using several use cases, two key fundamentally different aspects—control and design—are introduced and illustrated. In addition, we show how the World Avatar can improve interoperability between heterogeneous data formats as well as software, and thus enable cross-domain applications in wider contexts. Moreover, we highlight how the Parallel World framework can consider different scenarios and hence facilitate time-dependent what-if scenario analysis. All use cases show how interoperability between multiple domains involved in the complex decarbonisation process can contribute to CO_2 abatement of digitalisation.

M. Q. Lim · O. Inderwildi · M. Kraft (✉)
CARES, Cambridge Centre for Advanced Research and Education in Singapore, 1 Create Way, CREATE Tower, #05-05, Singapore 138602, Singapore
e-mail: mk306@cam.ac.uk

M. Q. Lim
e-mail: mei.qi.lim@cares.cam.ac.uk

O. Inderwildi
e-mail: oliver.inderwildi@scnat.ch

X. Wang
Department of Chemical and Biomolecular Engineering, National University of Singapore, 4 Engineering Drive 4, Singapore 117585, Singapore
e-mail: chewxia@nus.edu.sg

M. Kraft
Department of Chemical Engineering and Biotechnology, University of Cambridge, Philippa Fawcett Drive, Cambridge CB3 0AS, UK

School of Chemical and Biomedical Engineering, Nanyang Technological University, 62 Nanyang Drive, Singapore 637459, Singapore

© Springer Nature Switzerland AG 2022
O. Inderwildi and M. Kraft (eds.), *Intelligent Decarbonisation*, Lecture Notes in Energy 86, https://doi.org/10.1007/978-3-030-86215-2_4

4.1 Introduction

Global warming has a drastic impact on the environment, health, the economy, biodiversity, infrastructure, food and water supplies etc. For instance, the World Health Organisation (WHO) reports that climate change is expected to cause approximately 250,000 additional deaths yearly between 2030 and 2050, and the direct damage to health is estimated to cost between USD 2–4 billion per annum by 2030 (World Health Organisation 2018). Carbon dioxide (CO_2) released due to fossil fuel combustion is reported as the overwhelming contributor to greenhouse gas (GHG) emissions. The National Oceanic and Atmospheric Administration (NOOA) reported that the global average atmospheric CO_2 in 2019 was 409.8 ppm—the highest in the past 800,000 years (National Oceanic and Atmospheric Administration 2020). Moreover, the annual rate of increase in atmospheric CO_2 over the past 60 years is also noted to be approximately 100 times faster than previous natural increases. It has been estimated that if the global energy demand continues to grow and be satisfied predominately with fossil fuels, atmospheric CO_2 is projected to rise beyond 900 ppm by 2100 (National Oceanic and Atmospheric Administration 2020).

In 2017, the electricity and heat generation sector contributed 41.4% (13,603.3 million tonnes) to overall CO_2 emissions due to fuel combustion (International Energy Agency 2019). Therefore, the decarbonisation of energy provision plays a pivotal role in managing global GHG emissions and thus mitigating global warming. Inderwildi et al. (2020) highlighted that the digitalisation of energy systems using cyber-physical systems (CPS) can alter the marginal abatement cost curve (MACC) towards higher CO_2 abatement potentials. This implies that CPS have the potential to increase efficiency of energy provision and industrial production, and hence possess great emissions reduction potential. This potential increases further when CPS are combined with Artificial Intelligence (AI).

To fully unleash this potential, addressing and overcoming the current challenge of low interoperability between multiple domains involved in the complex decarbonisation process is a prerequisite. The challenge is attributed to the presence of information silos, e.g. disconnected data lakes and AIs arising from heterogeneity of data and services (software and tools), and the non-uniqueness of data i.e. data duplication and inconsistency. Any complex process involves collating large amounts of information, tools and models from multiple domains, which may comprise varying degrees of model resolution that are subjected to complex interaction loops. These information, tools and models typically have features of syntax and semantic heterogeneity, making communication across domains' boundaries challenging. Consequently, a substantial amount of effort is commonly required to process and convert the different data formats, and to navigate through a dynamic information base for decision-making. This results in the process being slow and labour-intensive while it does not make the best use of the available data, which in turn narrows its potential.

Doan et al. (2012) described how semantic data heterogeneity can be a major bottleneck for data integration. The semantic web and its technologies, such as Resource

Description Framework (RDF) and Web Ontology Language (OWL), offer the potential to increase the interoperability by formalising the representation of data and services. Semantic web technologies are standardised, provide a uniform way to query and link data, and have been approved for many years. In addition to being a suitable candidate for data integration, semantic web technologies provide unique features of knowledge management and reasoning in comparison to other approaches. This approach enables data and services that are often siloed at present to understand and communicate with each other, and thus use each other's functionalities. In particular, the ontology-based knowledge graphs demonstrate good prospects with the emergence of the first practical implementations in the context of the "World Avatar" project (Eibeck et al. 2019, 2020; Zhou et al. 2019, 2020; Inderwildi et al. 2020).

The purpose of this chapter is to:

- Introduce the concepts of the World Avatar project which is based on dynamic ontology-based knowledge graph
- Describe use cases that demonstrate the application of these concepts, emphasising two fundamentally different aspects—control and design.

4.2   The World Avatar—A Dynamic knowledge Graph

A knowledge graph represents information by making use of the principles of Linked Data as employed in the semantic web, where concepts correspond to vertices and relationships between concepts correspond to edges of the graph. As a representation of information, the key distinguishing feature of a knowledge graph is that individual aspects of the information are linked to each other. In the World Avatar project, this representation is implemented by means of ontologies, which formalise the definition of concepts and their relationships through collections of subject-predicate-object triples. The World Avatar concept intends to capture the idea of representing every aspect of the real world in a digital "mirror" world. This is essentially an extension of the Digital Twin notion, where, taking an example from Industry 4.0, a device or a unit operation in an industrial process has a corresponding virtual representation. A natural logical continuation is the application of this approach beyond the industrial context—the virtualisation of any abstract concept or process, similar to the extension of Internet of Things to the Internet of Services and beyond.

The "J-Park Simulator" (JPS)[1] (Eibeck et al. 2019) is an implementation and subset of the World Avatar concept. Figure 4.1 illustrates its main underlying principles. The World Avatar started with a focus on virtualising industrial operations within the Jurong Island Eco-Industrial Park (EIP) in Singapore (Pan et al. 2015, 2016; Zhou et al. 2017; Kleinelanghorst et al. 2017) using the concepts mentioned in Kraft and Mosbach (2010), but has since expanded well beyond this original scope. Fundamental to the World Avatar is a dynamic knowledge graph (dKG) that is envisioned as general-purpose and all-encompassing. In the World Avatar, the Linked

---

[1] http://www.theworldavatar.com/.

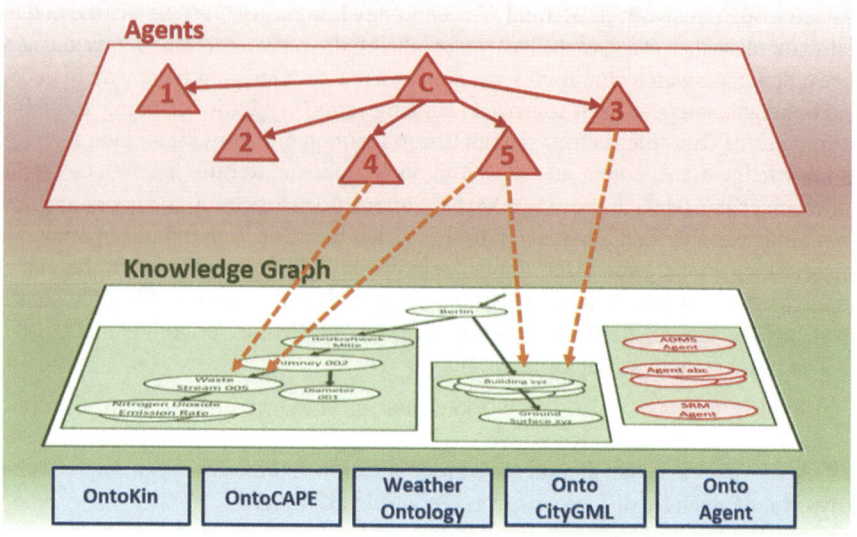

Fig. 4.1 An illustration of the main principles for the dynamic knowledge graph: (a) modular domain ontologies (blue), (b) instances of various types of agents (atomic, composition and composite) (red node), and (c) active agents (red triangles) operating on the knowledge graph and interacting with each other. "Agent" refers to software, methods, applications and services etc. that utilise semantic web technologies and operate on the knowledge graph to read/write, estimate, simulate, optimise and/or query etc. to fulfil specific objectives. Reproduced from Eibeck et al. (2019)

Data concept is implemented using Internationalised Resource Identifiers (IRIs),[2] a protocol element that complements Uniform Resource Identifiers (URIs). Essentially, IRIs are generalised web addresses that are used to point to a resource on the Web (World Wide Web Consortium 2008). Various modular domain ontologies have been employed in the World Avatar. These include OntoCAPE (for computer-aided process engineering Marquardt et al. 2010), OntoEIP (for EIPs Zhang et al. 2017; Zhou et al. 2017, 2018), OntoPowSys (for power systems Devanand et al. 2020) and OntoCityGML (for 3D models of cities and landscapes Eibeck et al. 2019). In the chemistry domain, ontologies were developed to semantically describe the subdomains of quantum chemistry calculations (OntoCompChem Krdzavac et al. 2019), species (OntoSpecies Farazi et al. 2020b), and chemical kinetic reaction mechanisms (OntoKin Farazi et al. 2020a). Furthermore, the use of Linked Data allows the World Avatar knowledge graph to connect to various subgraphs of the Linked Open Data (LOD) Cloud,[3] in particular DBpedia[4] (Lehmann et al. 2015), and leverage the wealth of data available on the Internet.

---

[2] https://www.w3.org/International/O-URL-and-ident.html.

[3] https://lod-cloud.net/.

[4] https://wiki.dbpedia.org/.

Beyond mere data representation, the World Avatar contains an ecosystem of software agents that act autonomously and continuously on the knowledge graph, constantly updating it and thus making it evolve in time. Crucially, the agents themselves are part of the knowledge graph, governed by an agent ontology (OntoAgent Zhou et al. 2019). In particular, agents were developed for automatic agent discovery and composition (Zhou et al. 2019), i.e. agents that create new, composite agents for more complex tasks. Furthermore, to facilitate the use of agents and simplify the identification of an agent suitable for a specific task in an agent-rich environment (where an abundance of services is available), an agent marketplace based on blockchain technology and Smart Contracts was established (Zhou et al. 2020).

With its generic, all-purpose design based on ontologies and autonomous agents, the World Avatar improves interoperability between heterogeneous data formats as well as software, and thus enables cross-domain applications in wider contexts. For example, the World Avatar has been employed in the optimal site selection for modular nuclear power plants (Devanand et al. 2019) and simulation of atmospheric dispersion of pollutants in emissions use cases involving power plants (Eibeck et al. 2019). In addition, the World Avatar provides the capability to consider different scenarios for scenario planning via the usage of the "Parallel World" framework (Eibeck et al. 2020).

4.3   Use Cases

In this section, several use cases that demonstrate the application of a dKG will be presented. These use cases are associated with Singapore and primarily with its Jurong Island EIP due to the above-mentioned reason. However, in principle, the application of the dKG is not restricted to these contexts—the dKG could be extended to, for instance, other countries and sectors.

The five use cases have been categorised into two groups to introduce and illustrate two fundamentally different aspects of the World Avatar concept—control and design. The first three use cases demonstrate how digital twinning in the context of a dKG can reduce costs and energy via intelligent control strategies. The subsequent two use cases exemplify how the Parallel World framework can be employed to create a "live" digital world i.e. a scenario that enables the investigation of different technology alternatives and the effect of policies on technology transition.

4.3.1   Digital Twinning—Intelligent Control Strategies

Singapore, an island city-state in Southeast Asia, is a modern city with highly developed infrastructure. Apart from being one of the world's most densely populated countries (The World Bank Group 2018), with a population of 5.69 million people (Singapore Department of Statistics 2020), Singapore also hosts one of the world's busiest ports and leading energy and chemicals hubs. Situated off the southern coast of Singapore, Jurong Island is the core of Singapore's energy and chemicals industry. With a land area of 31 square kilometres, Jurong Island is home to over 100 global petroleum, petrochemical and specialty chemical companies (Singapore Economic Development Board 2018) and has attracted over S\$50 billion worth of investments (Singapore Economic Development Board 2020). Some of these companies such as Evonik, BASF, ExxonMobil, Linde, Shell etc. are prominent players in the field. In 2014, with 1.5 million barrels of oil being refined per day on Jurong Island, Singapore is one of the top ten exporters of refined oil products in Asia (Singapore Economic Development Board 2020). In 2015, the energy and chemicals industry also contributed S\$81 billion—about a third of Singapore's total manufacturing output (Singapore Economic Development Board 2020). These figures illustrate the importance of the energy and chemicals industry to Singapore's economy. Unfortunately, the energy and chemicals industry is also the largest source of emissions in Singapore.

As mentioned in Sect. 4.2, the World Avatar project started with a focus on virtualising industrial operations within Jurong Island. In the following subsections, three use cases which comprise multiple domains such as energy, electrical and chemical networks, with varying scales and degrees of detail, will demonstrate how digital twinning and intelligent control strategies in the context of a dKG can increase efficiency of energy provision and CO_2 abatement potentials.

4.3.1.1 Control Strategy—Depropaniser Section

Although chemical processes and electrical power system operations are usually analysed separately, in reality, any change in the chemical processes will be reflected in the corresponding electrical load demand profile and might affect the transient stability and power quality of the electrical system. In this use case, the dKG is utilised to recommend the control strategy for the feed flow rate of the depropaniser section within a typical natural gas processing plant. The relevant chemical and electrical aspects have been modelled using the software gPROMS[5] and MATLAB[6] respectively. The two programs have been packaged as two distinct agents that can apply the semantic web stack to read and understand information from the dKG and modify its data values. The real power consumed by the chemical processes is used to couple the chemical and electrical systems. The dKG establishes machine-to-machine (M2M) communications between the two typically siloed systems via an agent framework, and ontology called "OntoTwin". The OntoTwin ontology is developed based on combining and extending the aforementioned OntoPowsys and OntoEIP ontologies. The agent framework employs Description Logics (DL), SPARQL Inferencing Notation[7] (SPIN) reasoning techniques, as well as detailed models of the selected chemical

---

[5] https://www.psenterprise.com/products/gproms.

[6] https://www.mathworks.com/products/matlab.html.

[7] https://spinrdf.org/.

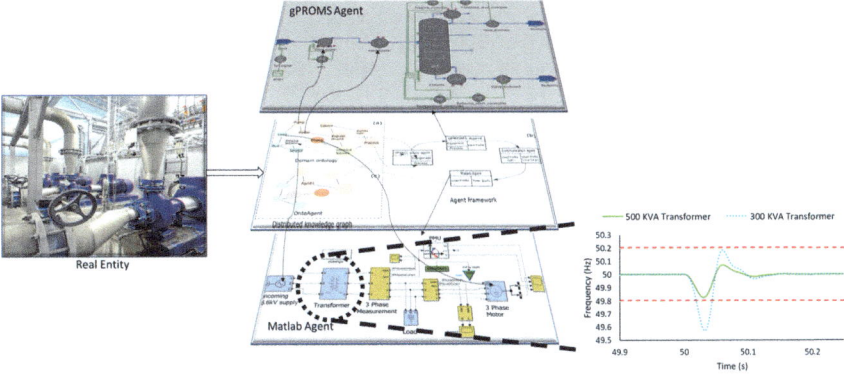

Fig. 4.2 Digital twin of the depropaniser section within a typical natural gas processing plant

and electrical systems (as shown in Fig. 4.2) to investigate the interactions between the two systems. As operating constraints such as product and power qualities have been incorporated into the framework, the framework is capable of predicting any constraint violation arising from the proposed control strategy—even before their implementations.

Moreover, a comparative study was conducted with two different ratings of transformer—300 and 500 kVA (standard design). The study revealed that by adopting the recommended control strategy, a lower-rated and thus cheaper transformer i.e. 300 kVA will be sufficient for the specified purpose. This was estimated to translate to a cost saving of approximately EUR 40,000. The use case illustrates how digital twinning (of the depropaniser section) in the context of a dKG can provide smooth operation and better design through: (1) maintaining both the product and power qualities within their desired operating ranges, thus reducing the quantity of off-spec products and downtime; and (2) the potential to utilise a lower-rated transformer attributed to the adoption of the recommended control strategy (Devanand et al. 2022).

4.3.1.2   Control Strategy—Heat and Power Dispatch

Combined Heat and Power (CHP) cogeneration system is an energy-efficient technology that concurrently produces electrical and useful thermal energy such as steam or hot water. In conventional electricity generation, approximately two-thirds of the energy used is dissipated in the form of heat to the atmosphere (United States Environmental Protection Agency 2019). By capturing and transforming this waste heat, CHP can attain energy efficiency of over 80% (United States Environmental Protection Agency 2019). Consequently, CHP is commonly deployed at sites that require both electrical and thermal energy e.g. Jurong Island. In this use case, the dKG is utilised to determine the optimal scheduling for the operation of the Combined

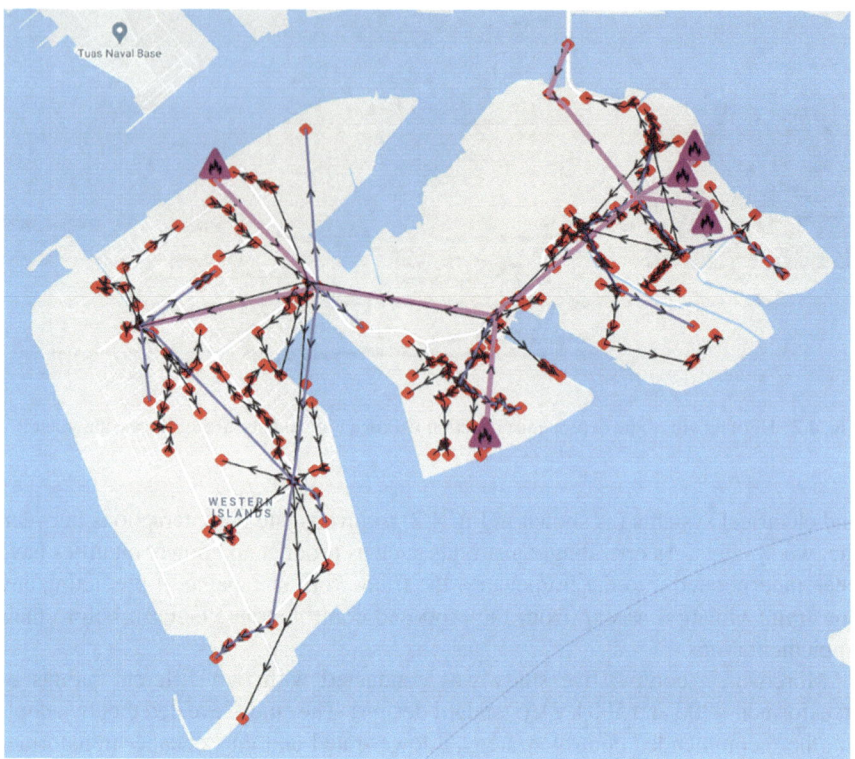

Fig. 4.3 Digital twin—transmission grid. The coloured lines represent the transmission lines in the electrical network that connect buses which are represented by the red points. Different coloured transmission lines represent different voltage levels: pink represents 230 kV, purple represents 66 kV and black represents 22 kV. The arrow indicates the direction of the current flow from one bus to another within the network. The gas symbols denote the power plants

Cycle Gas Turbines (CCGTs) within a grid. The objective is to propose heat and power dispatch to fulfil both the thermal and electrical loads, as well as grid constraints, while minimising the overall CO_2 emissions. A detailed digital twin of an electrical network has been developed and described with the OntoPowSys ontology (Devanand et al. 2020) for this purpose. Figure 4.3 visualises a subset of this digital twin—representation of the transmission grid. Comparing with uncontrollable cogeneration units i.e. priority is given to electrical energy, it is concluded that significant CO_2 emissions reduction can be attained via the efficient power dispatch strategy, in particular for higher heat load levels. More details on the implementation and results can be found in Rigo-Mariani et al. (2019, 2020).

4.3.1.3 Control Strategy—Energy Demand Side Management

Distributed energy resources (DERs) are technologies that primarily consist of small, modular energy generation and storage systems. DERs may either be connected to the local electric power grid or stand-alone applications (National Renewable Energy Laboratory 2002). With the growing attention on renewable energy, generation technologies such as wind turbines and photovoltaics are also being incorporated into the DERs. In the meantime, technology advancements on the energy demand side will result in the automation of smart controllable loads and the growth in demand side management schemes such as demand response (Li et al. 2019).

Building upon the work of Li et al. (2019), an energy demand side management framework was developed in the context of a dKG. The objective of this agent framework is to reduce the peak to average ratio, which in turn leads to lower stress on the main grid and hence provides cost savings for the consumers. The current agent framework considers the forecasted renewable energy generated by solar photovoltaic cells, three types of energy consumers (residential, commercial and industrial) as well as their demand flexibilities. The dKG establishes communications between the various players and data via the OntoPowSys ontology and an agent framework that employs a game theory modelling approach. Furthermore, energy trading based on the optimised energy profiles is secured via the use of blockchain technology. This use case illustrates how digital twinning (of the players) in the context of a dKG can: (1) reduce the overall CO_2 emissions by increasing penetration of renewable energy sources while fulfilling the players' demand constraints; and (2) through the application of blockchain technology, create a seamless, secured and efficient energy system that lowers the threshold of participation in the electricity market for small players.

4.3.2 Digital Twinning—Intelligent Design Strategies

In a complex decision environment, decision-making processes face the challenge of predicting and/or estimating the impact of any decision. Scenario analysis facilitates decision-making by exploring and presenting different options and their corresponding outcomes and implications. As mentioned in Sect. 4.2, the dKG provides the capability to consider different scenarios via the usage of the Parallel World framework. The underlying knowledge-graph-based design of the World Avatar project and semantic web technologies allow instances (from different sources) within the dKG to be annotated with additional information (e.g. versioning and provenance data). This corresponds to the "live" Parallel World concept which keeps entities involved in the same scenario together and delegates their access, queries and updates on the dKG to a scenario-specific portion of the dKG. The scenario-specific portion only overshadows the portion of the dKG where modifications are necessary. Entities that are unchanged remain connected to the "Base World" of the dKG. This architecture allows entities involved in the same scenario to: (1) maintain connections to the Base

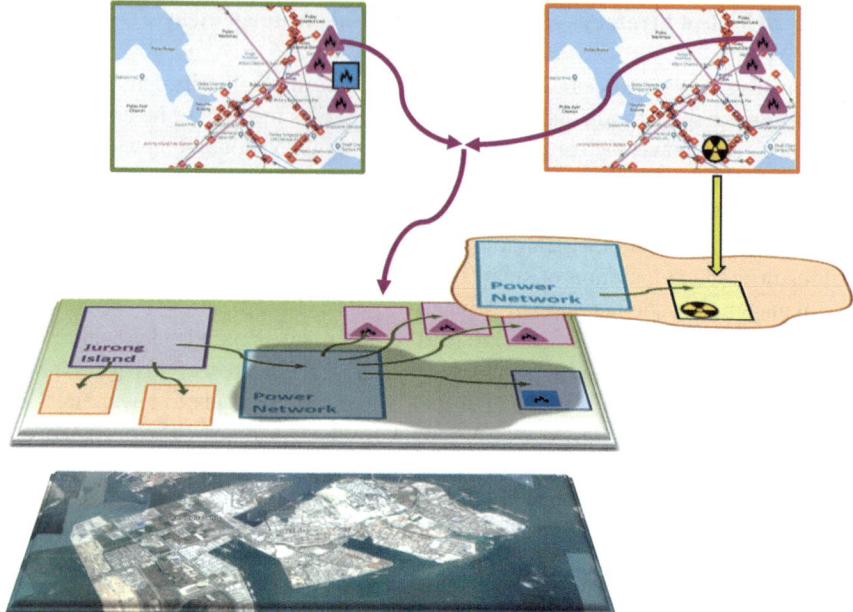

Fig. 4.4 Parallel World concept for what-if scenario analysis. Entities are represented and linked semantically in the knowledge graph. Reproduced from Eibeck et al. (2020)

World i.e. modifications in the Base World will be reflected in the Parallel Worlds; and (2) operate in a "secured portion" of the dKG without interfering with other entities or impacting the Base World.

The Parallel World framework also enables data to be linked and queried in a highly flexible manner. Consequently, the Parallel World concept can assist in design space exploration and its management by allowing the modification of relevant data and parameters, and storing the different versions and scenario results in the dKG without mutual interference (i.e. where each world is configured with its own data and describes a potential state). More details on the Parallel World framework and its implementation can be found in Eibeck et al. (2020) (Fig. 4.4).

4.3.2.1 Design Strategy—Energy Storage Technology Selection and Placement

One of the major challenges for increasing the penetration of renewable energy sources is their availability and intermittency, which can be addressed with the deployment of energy storage systems (ESSs). ESS involves converting excess energy to a form that can be stored e.g. electrochemical (lithium-ion battery) or mechanical (pumped hydro), and transforming the stored energy back into electricity when it is required. Driven by the growing attention on ESS, many types of ESS and applications have emerged during the last few decades. Consequently, a systematic framework that facilitates the selection of suitable energy storage technologies among the increasing options is necessary (Li et al. 2018).

Building upon the work of Li et al. (2018), an energy storage technology selection and placement framework was developed in the context of a dKG. The objective of the framework is to recommend appropriate energy storage technology based on a variety of factors such as technological feasibility, maturity, installation cost, etc. Once the energy storage technology is determined, the optimal placement of the ESS for a given grid topology to minimise distribution system losses is established based on the system loss sensitivity index with respect to the ESS parameters. The current agent framework considers six types of energy storage technologies and has been applied to the aforementioned digital twin of an electrical network (Sect. 4.3.1.2). The OntoPowSys ontology was extended to describe and instantiate the relevant energy storage technologies in the dKG. This use case illustrates how digital twinning (of the electrical network and different energy storage technologies) in the context of a dKG can: (1) recommend optimal ESS selection and placement strategy; and (2) reduce the overall CO_2 emissions by encouraging penetration of renewable energy sources.

4.3.2.2   Design Strategy—Effect of Carbon Tax on Technology Transition

A carbon tax is a tax levied on the combustion of fossil fuels e.g. coal, oil and natural gas, and is associated with the global warming potentials of its emissions. The carbon tax aims to motivate the transition to alternative clean energy by discouraging the use of fossil fuels whose combustion generates GHG (primarily CO_2). In this use case, the dKG is utilised to address the following questions: (1) For a set of conditions e.g. a generator's characteristics (design capacity, carbon emission factor, capital cost, fuel cost etc.), and a project's life span, depreciation rate and load profiles, what is the minimum value of carbon tax required to motivate (i.e. it becomes profitable) the replacement of existing power plant(s) with small modular nuclear reactors (SMRs); (2) which existing power plant(s)/generator(s) need to be replaced; (3) what are the required number and optimal locations for the new SMRs; and (4) how to connect these SMRs to the existing transmission grid. The current agent framework has been applied to the aforementioned digital twin of an electrical network (Sect. 4.3.1.2) which comprises six oil and 22 natural gas generators. Similarly, the OntoPowSys ontology was extended to describe and instantiate the relevant SMRs and generators in the dKG. When the carbon tax is increased from $5 to $170, the oil generators are replaced with SMRs in the Parallel World. The corresponding estimated CO_2 emissions and types of generators are updated automatically in the dKG to reflect these changes. Subsequently, an Optimal Power Flow (OPF) agent is invoked to minimise the overall operating cost. More details on the implementation and results can be found in Devanand et al. (2019) and Eibeck et al. (2020). This use case

Fig. 4.5 The illustration on the left depicts the original electrical network and the illustration on the right depicts the modified electrical network. The blue square denotes an oil generator, pink triangles denote natural gas generators, the and radiation symbol denotes a small modular nuclear reactor

illustrates how digital twinning in the context of a dKG can allow decision-makers to better understand the effect of policy instruments such as carbon tax and therefore helps to move faster toward a low-carbon economy (Fig. 4.5).

4.4 Outlook and Conclusion

The main underlying concepts and principles of the World Avatar project—which is based on dKG—have been described in this chapter. Using several use cases, two key fundamentally different aspects, control and design, have also been introduced and illustrated. In addition, it was discussed how the World Avatar can improve interoperability between heterogeneous data formats as well as software, and thus enables cross-domain applications in wider contexts. Moreover, it was highlighted how the Parallel World framework can consider different scenarios and hence facilitate what-if scenario analysis.

Despite the above-mentioned work, there are many open questions to be answered and more work to be done to advance and further leverage the potential of the World Avatar concepts. Some of these areas are described as follows:

1. Extend the dKG with additional domains and their corresponding ontologies e.g. financial, health and safety, and new instances e.g. digital twin of the European and United Kingdom electrical network.
2. Create and integrate more software agents for various application domains into the dKG. This includes the use of machine learning algorithms to increase computational speed and intelligence.

3. Increase the dKG's connectivity and application to real/physical systems e.g. via sensors and actuators. This includes the use of satellite data and other publicly available data sources.
4. Resolve conflicting and incomplete domain ontologies e.g. through ontology matching techniques employing natural language processing and machine learning.
5. Improve the current dKG's query, access and writing speed e.g. via the use of indexing.
6. Enhance the capability, performance and scalability of the semantic agent composition framework to facilitate automatic agent discovery and composition.
7. Include cyber security and resilience aspects in the dKG.
8. Introducing goals that guarantee human-friendly operation and design, including an ontology for the UN sustainable development goals, has been identified as a first step in this direction.

Despite this long list of necessary improvements we conclude that the World Avatar project has the potential to address and overcome the challenge of low interoperability between multiple domains involved in the complex decarbonisation process, therefore realising the CO_2 abatement potential of digitalisation.

Acknowledgements This research is supported by the National Research Foundation, Prime Minister's Office, Singapore under its Campus for Research Excellence and Technological Enterprise (CREATE) programme. MK gratefully acknowledges the support of the Alexander von Humboldt foundation.

References

Devanand A, Karimi IA, Kraft M (2019) Optimal site selection for modular nuclear power plants. Comput Chem Eng 125:339–350. https://doi.org/10.1016/j.compchemeng.2019.03.024

Devanand A, Karmakar G, Krdzavac N, Rigo-Mariani R, Foo YS, Karimi IA, Kraft M (2020) OntoPowSys: a power system ontology for cross domain interactions in an eco industrial park. Energy AI 1:100008. https://doi.org/10.1016/j.egyai.2020.100008

Devanand A, Karmakar G, Krdzavac N, Farazi F, Lim MQ, Foo Eddy YS, Karimi IA, Kraft M (2022) ElChemo: A cross-domain interoperability between chemical and electrical systems in a plant. Comput Chem Eng 156, 107556. https://doi.org/10.1016/j.compchemeng.2021.107556

Doan A, Halevy A, Ives Z (2012) Principles of data integration. Morgan Kaufmann

Eibeck A, Lim MQ, Kraft M (2019) J-Park simulator: an ontology-based platform for cross-domain scenarios in process industry. Comput Chem Eng 131:106586. https://doi.org/10.1016/j.compchemeng.2019.106586

Eibeck A, Chadzynski A, Lim MQ, Aditya K, Ong L, Devanand A, Karmakar G, Mosbach S, Lau R, Karimi IA et al (2020) A parallel world framework for scenario analysis in knowledge graphs. Data-Centric Eng 1:e6. https://doi.org/10.1017/dce.2020.6

Farazi F, Akroyd J, Mosbach S, Buerger P, Nurkowski D, Salamanca M, Kraft M (2020a) OntoKin: an ontology for chemical kinetic reaction mechanisms. J Chem Inf Model 60(1):108–120. https://doi.org/10.1021/acs.jcim.9b00960

Farazi F, Krdzavac N, Akroyd J, Mosbach S, Menon A, Nurkowski D, Kraft M (2020b) Linking reaction mechanisms and quantum chemistry: an ontological approach. Comput Chem Eng 137:106813. https://doi.org/10.1016/j.compchemeng.2020.106813

Inderwildi O, Zhang C, Wang X, Kraft M (2020) The impact of intelligent cyber-physical systems on the decarbonization of energy. Energy Environ Sci 13(3):744–771. https://doi.org/10.1039/C9EE01919G

International Energy Agency (2019) CO_2 emissions from fuel combustion highlights. http://www.shorturl.at/cpDW3. Accessed 22 October 2020

Kleinelanghorst MJ, Zhou L, Sikorski J, Foo YS, Aditya LK, Mosbach S, Karimi IA, Lau R, Kraft M (2017) J-Park simulator: roadmap to smart eco-industrial parks. In: Proceedings of the second international conference on internet of things, data and cloud computing (ICC'17), pp 1–10. https://doi.org/10.1145/3018896.3025155

Kraft M, Mosbach S (2010) The future of computational modelling in reaction engineering. Philos Trans 368:3633–3644. https://doi.org/10.1098/rsta.2010.0124

Krdzavac N, Mosbach S, Nurkowski D, Buerger P, Akroyd J, Martin J, Menon A, Kraft M (2019) An ontology and semantic web service for quantum chemistry calculations. J Chem Inf Model 59(7):3154–3165. https://doi.org/10.1021/acs.jcim.9b00227

Lehmann J, Isele R, Jakob M, Jentzsch A, Kontokostas D, Mendes PN, Hellmann S, Morsey M, van Kleef P, Auer S, Bizer C (2015) DBpedia: a large-scale, multilingual knowledge base extracted from Wikipedia. Semantic Web 6(2):167–195. https://doi.org/10.3233/SW-140134

Li L, Liu P, Li Z, Wang X (2018) A multi-objective optimization approach for selection of energy storage systems. Comput Chem Eng 115:213–225. ISSN 0098-1354. https://doi.org/10.1016/j.compchemeng.2018.04.014

Li Y, Yang W, He P, Chen C, Wang X (2019) Design and management of a distributed hybrid energy system through smart contract and blockchain. Appl Energy 248:390–405. ISSN 0306-2619. https://doi.org/10.1016/j.apenergy.2019.04.132

Marquardt W, Morbach J, Wiesner A, Yang A (2010) OntoCAPE: a re-usable ontology for chemical process engineering, 1st ed. Springer-Verlag, Berlin, Heidelberg. https://doi.org/10.1007/978-3-642-04655-1

National Oceanic and Atmospheric Administration (2020) Climate change: atmospheric carbon dioxide. http://www.shorturl.at/lmqsH. Accessed 22 October 2020

National Renewable Energy Laboratory (2002) Using distributed energy resources. https://www.nrel.gov/docs/fy02osti/31570.pdf. Accessed 22 October 2020

Pan M, Sikorski J, Kastner CA, Akroyd J, Mosbach S, Lau R, Kraft M (2015) Applying Industry 4.0 to the Jurong Island eco-industrial park. Energy Proc 75:1536–1541. https://doi.org/10.1016/j.egypro.2015.07.313

Pan M, Sikorski J, Akroyd J, Mosbach S, Lau R, Kraft M (2016) Design technologies for eco-industrial parks: From unit operations to processes, plants and industrial networks. Appl Energy 175:305–323. https://doi.org/10.1016/j.apenergy.2016.05.019

Rigo-Mariani R, Ling KV, Maciejowski J (2019) A clusterized energy management with linearized losses in the presence of multiple types of distributed generation. Int J Electrical Power Energy Syst 113:9–22. ISSN 0142-0615. https://doi.org/10.1016/j.ijepes.2019.04.049

Rigo-Mariani R, Zhang C, Romagnoli A, Kraft M, Ling KV, Maciejowski J (2020) A combined cycle gas turbine model for heat and power dispatch subject to grid constraints. IEEE Trans Sustain Energy 11(1):448–456. https://doi.org/10.1109/TSTE.2019.2894793

Singapore Department of Statistics (2020) Singstat. https://www.singstat.gov.sg/. Accessed 22 October 2020

Singapore Economic Development Board (2018) Singapore: a leading manufacturing hub. https://www.edb.gov.sg/en/news-andevents/insights/innovation/singapore-a-leading-manufacturing-hub.html. Accessed 22 October 2020

Singapore Economic Development Board (EDB) (2020) The right chemistry. https://www.edb.gov.sg/en/our-industries/energy-and-chemicals.html. Accessed 22 October 2020

The World Bank Group (2018) Population density (people per sq. km of land area). https://data.worldbank.org/indicator/EN.POP.DNST. Accessed 22 October 2020

United States Environmental Protection Agency (2019) Combined heat and power (CHP) partnership. https://www.epa.gov/chp/what-chp. Accessed 22 October 2020

World Health Organisation (2018) Climate change and health. https://www.who.int/news-room/fact-sheets/detail/climate-change-and-health. Accessed 22 October 2020

World Wide Web Consortium (2008) An introduction to multilingual web addresses. https://www.w3.org/International/articles/idn-and-iri/. Accessed 22 October 2020

Zhang C, Romagnoli A, Zhou L, Kraft M (2017) Knowledge management of eco-industrial park for efficient energy utilization through ontology-based approach. Appl Energy 204:1412–1421. https://doi.org/10.1016/j.apenergy.2017.03.130

Zhou L, Pan M, Sikorski JJ, Garud S, Aditya LK, Kleinelanghorst MJ, Karimi IA, Kraft M (2017) Towards an ontological infrastructure for chemical process simulation and optimization in the context of eco-industrial parks. Appl Energy 204:1284–1298. https://doi.org/10.1016/j.apenergy.2017.05.002

Zhou L, Zhang C, Karimi IA, Kraft M (2018) An ontology framework towards decentralized information management for eco-industrial parks. Comput Chem Eng 118:49–63. https://doi.org/10.1016/j.compchemeng.2018.07.010

Zhou X, Eibeck A, Lim MQ, Krdzavac N, Kraft M (2019) An agent composition framework for the J-Park simulator: a knowledge graph for the process industry. Comput Chem Eng 130:106577. https://doi.org/10.1016/j.compchemeng.2019.106577

Zhou X, Lim MQ, Kraft M (2020) A smart contract-based agent marketplace for the J-Park simulator: a knowledge graph for the process industry. Comput Chem Eng 139:106896. https://doi.org/10.1016/j.compchemeng.2020.106896

Chapter 5
Insights: AI and Decarbonisation

David Rolnick

1. What roles can AI play in reducing CO_2 emissions?

AI can refer to any computer algorithm that performs a complicated task, including machine learning algorithms that "learn" from data. These methods are not magic, but they are good at analysing large amounts of data to pick out patterns, and at scaling up simple tasks that humans could do more slowly. There are several roles that AI can take in accelerating decarbonisation. AI can help distil large amounts of information to help policymakers make decisions—for example, by using satellite imagery to automatically monitor greenhouse gas emissions or land use. AI can help optimise complicated systems—for example, improving the efficiency of freight transportation or controlling the heating and cooling system of a building to reduce the energy consumed. AI can improve forecasting—for example, predicting prices in carbon markets or power generation from solar and wind to help balance the electrical grid. AI can also accelerate the process of scientific experimentation, for example by suggesting promising candidate materials for technologies such as batteries and photovoltaics.

2. Are there pitfalls to be aware of?

It's easy to be overly optimistic about a shiny, new technology. AI is a powerful tool, but it isn't a silver bullet, and fancy algorithms are only useful in certain circumstances. Also, like any other technology, AI is not intrinsically positive or negative, and it can be used to increase CO_2 emissions as well as decrease them—in particular, there are many applications of AI in fossil fuel exploration and extraction. AI algorithms also consume energy directly, though this is arguably a small effect compared to the positive or negative impacts of the applications for which these algorithms are used. Finally, it is vital for applications of AI to be developed so as to empower

D. Rolnick (✉)
Assistant Professor of Computer Science, McGill University, Montreal, Canada
e-mail: drolnick@mila.quebec

© Springer Nature Switzerland AG 2022
O. Inderwildi and M. Kraft (eds.), *Intelligent Decarbonisation*, Lecture Notes in Energy 86, https://doi.org/10.1007/978-3-030-86215-2_5

individuals, communities and nations across the world, rather than consolidating technological power and perpetuating a "digital divide".

3.   What is needed to enable impactful work in AI for decarbonisation?

The biggest gaps are in understanding and communication. It is important to facilitate multi-stakeholder work that brings together complementary sets of expertise, and to build capacity in AI within public and private entities, so as to facilitate impactful applications well-tailored to decarbonisation challenges. It is also essential to develop standards for gathering, sharing and using data in this space that are sensitive to privacy and the needs of society.

4.   Can you tell us about Climate Change AI?

Climate Change AI is a global organisation of experts from academia, industry and policy who work at the intersection of AI and climate change. We provide resources such as grants, course materials and tutorials; run conferences and other events for knowledge-sharing across sectors; and advise stakeholders such as policymakers who want to understand the opportunities and pitfalls that exist in this space. I encourage those who are interested in this topic to explore the resources at www.climatech ange.ai, including our 100-page report "Tackling Climate Change with Machine Learning" detailing opportunities that exist within this space.

David Rolnick is Assistant Professor and Canada CIFAR AI Chair in the School of Computer Science at McGill University and at the Mila Quebec AI Institute. He is co-founder and chair of Climate Change AI and serves as scientific co-director of Sustainability in the Digital Age. Dr. Rolnick has also worked at Google and DeepMind, and is a former NSF Mathematical Sciences Postdoctoral Research Fellow, NSF Graduate Research Fellow and Fulbright Scholar. He received his Ph.D. in Applied Mathematics from MIT.

Chapter 6
Insights: Intelligent Decarbonisation in Singapore

Teck Hua Ho

1.   What are the main goals of AI Singapore?

AI Singapore (AISG) was launched in June 2017 as an integrated, impact-driven research and innovation programme in artificial intelligence (AI) for Singapore. As a national initiative, AISG brings together the research strengths of Singapore's autonomous universities and research institutes, as well as the vibrant local ecosystem of AI start-ups and companies developing AI products, to perform use-inspired research, create innovative AI solutions and develop local talent to power Singapore's AI efforts.

2.   What role can AI play in the reduction of CO_2 emissions today?

AI has the potential to play a big role in the reduction of CO_2 emissions. AI-powered solutions, for example, could help policy-makers, business owners and industry leaders manage greenhouse gas emissions and support sustainability initiatives. In 2020, PwC and Microsoft predicted that the use of AI could reduce worldwide greenhouse gas emissions by 4% by 2030—an amount equivalent to 2.4 gigatonnes of CO_2 emissions—through various process optimisation efforts. Some other areas where AI can play a positive role are agriculture, energy, transportation and manufacturing, through a combination of improved monitoring, testing and design optimisation.

At AISG, we will be applying AI at our new HPC-AI Data Centre (projected to be ready near the end of 2021) to optimise operations and lower running costs. Optimising HPC-AI tasks to improve server loads will help us better manage air conditioner usage and electrical consumption.

T. H. Ho (✉)
National University of Singapore (NUS), Singapore, Singapore
e-mail: teck@nus.edu.sg

© Springer Nature Switzerland AG 2022
O. Inderwildi and M. Kraft (eds.), *Intelligent Decarbonisation*, Lecture Notes in Energy 86, https://doi.org/10.1007/978-3-030-86215-2_6

3. Are there any wider challenges in using AI technology?

There are a few pressing issues. At present, AI-based systems are computationally intensive and rely heavily on the use of servers and data centres, indirectly adding to the carbon footprint. Research is being done to optimise energy consumption at data centres via revolutionary AI learning techniques such as a one-shot learning model, transfer learning, AI learning from small dataset and edge AI.

More recently, society has started to grapple with just how much human biases can make their way into AI systems and solutions, often with harmful results. In order to facilitate rapid deployment and increase the uptake of AI solutions by industry, being able to detect, reduce and remove bias from AI solutions is a key priority.

Last but not least, we also need to increase public awareness and the understanding of AI, its limitations and how it can transform and benefit people's lives. The biggest misconception may be that in the near future, we will achieve a technological singularity and create or help create a super AI that may be hard to control, and that will ultimately control us. The truth is that this is very far from reality. At present, AI is only very good at accomplishing specific, well-defined tasks. While it is true that AI will take over some jobs that people commonly do, this only applies to jobs that are predefined, narrow and repetitive, or those that can be outsourced. It is important to note that AI is unlikely to take over jobs that require creativity and empathy. And perhaps most importantly, AI will likely create many new jobs, and new types of jobs and job opportunities.

4. What technical advances in AI technology can we expect in the near and not so near future and what does this mean for sustainable living?

AI will be more and more pervasive and embedded in our lives, probably to the point that we will not even realise that we are using it. Many of these applications are geared towards automation, improving efficiency and productivity and optimising resources—very much in alignment with sustainability. For this to happen, some if not all of the above-mentioned challenges need to be addressed. We have already seen some of the initiatives bearing fruit—in January 2020, for example, the Singapore government launched their second edition of the Model AI Governance Framework. The framework provides detailed and readily implementable guidance to private sector organisations to address key ethical and governance issues when deploying AI solutions. By explaining how AI systems work, building good data accountability practices, and creating open and transparent communication, the framework aims to promote the public understanding of and trust in AI technologies.

At AI Singapore, we are actively researching bias, transparency and the explainability of AI solutions, as well as advanced AI that learns through one-shot learning and with a small dataset. We are also pushing for applications and the adoption of AI solutions in healthcare, logistics, manufacturing, professional services and education. This is in addition to the talent development and outreach programmes that we offer to industry and the general public, to prepare our citizens to be AI-enabled, AI ready and AI-aware.

Teck-Hua Ho is the senior deputy president and provost at the National University of Singapore (NUS), where he is a Tan Chin Tuan Centennial Professor. He is also the executive chairman of AI Singapore, a national research and development programme, and chairman of the Singapore Data Science Consortium. Teck is a prominent behavioural scientist with a PhD in decision sciences from the Wharton School of the University of Pennsylvania. He has a bachelor's degree in electrical engineering with first-class honours, a master's degree in computer and information sciences from NUS, and a master's degree in decision sciences from the Wharton School.

Chapter 7
Blockchain for Decarbonization

Choh Yun Bin, Wentao Yang, and Xiaonan Wang

Abstract Industry 4.0 has ushered in a new era of connectivity and communication within various industries. Motivated by the United Nation's Sustainable Development Goals (SDGs) of achieving net zero emissions by 2050, long-term energy sustainability plans encourage decentralized/distributed technologies such as blockchain to take center stage, alongside Internet of Things (IoT) devices, smart sensors, and smart contracts. Blockchain is posited to be impactful in decarbonizing various industries as it is an immutable, secure, and transparent ledger that incentivizes industrial adoption through reduced costs and increased efficiency. Therefore, in this book chapter, we will delve into the motivation of blockchain applications in the following industries: 1) energy markets, 2) chemical and manufacturing industries, and 3) carbon trading markets. Various case studies and current blockchain practices in each respective industry will also be reviewed and we will discuss the potential of blockchain towards decarbonization.

7.1 Introduction

Energy markets across the globe are being revolutionized rapidly to meet the rising demands of renewable energy sources (RES). Over the last decade, renewable technologies have increased global generation in the energy mix from 5.9% in 2009 to 13.4% in 2019 (Centre 2020). Coupled with a decrease in the levelized cost of electricity (LCOE), the path towards global decarbonization, as stated in the Paris Agreement, has never been brighter. This sharp inclination towards renewables

C. Y. Bin
National University of Singapore, Singapore 117585, Singapore

W. Yang
School of Economics and Management, Tsinghua University, Beijing 100084, China

China Zhigui Internet Technology, Yinke Building, Beijing 100080, China

X. Wang (✉)
Tsinghua University, Beijing 100084, China
e-mail: wangxiaonan@tsinghua.edu.cn

© Springer Nature Switzerland AG 2022
O. Inderwildi and M. Kraft (eds.), *Intelligent Decarbonisation*, Lecture Notes in Energy 86, https://doi.org/10.1007/978-3-030-86215-2_7

and Distributed Energy Resources (DERs) such as solar photovoltaics (PV), wind and energy storage has tipped the balance of power from the traditional/centralized authorities to the direct consumers of the grid. Therefore, distributed ledger technologies (DLT) such as blockchain, are the catalysts needed to speed up the progress in such distributed energy systems.

Blockchain went mainstream after Satoshi Nakamoto adapted the technology in 2009 to create the digital cryptocurrency Bitcoin. The main idea behind blockchain, is to establish a trust system between users across a shared network. Blockchain is a near real-time, distributed, and immutable ledger that operates on a peer-to-peer (P2P) basis. This technology eliminates the need for a central intermediary, thus saving time and cost, while providing a high level of transparency and security. To ensure a synchronized distribution of this ledger across all nodes in the network, blockchain validation operates through a distributed consensus algorithm (see Fig. 7.1).

To generate and add a block onto the existing chain, the block must be accepted by the network members. The consensus is achieved through algorithms such as Bitcoin's famous 'HashCash' Proof of Work (PoW) algorithm. However, not all blockchains use the PoW algorithm due to its unsustainable high energy consumption caused by the intensive computational power required to run the PoW algorithm. For example, in the energy market, Proof of Stake (PoS) and Proof of Authority (PoA) algorithms could be more well suited due to the exchange of sensitive data in

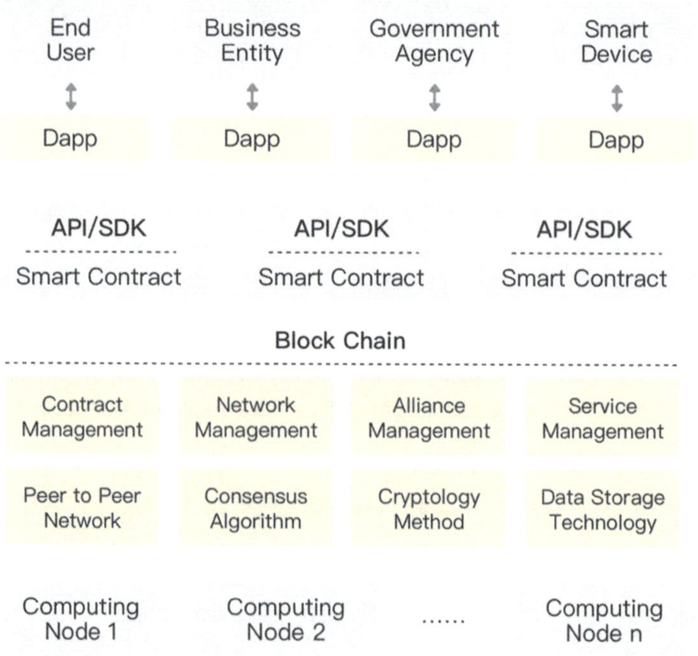

End User	Business Entity	Government Agency	Smart Device
↕	↕	↕	↕
Dapp	Dapp	Dapp	Dapp

API/SDK	API/SDK	API/SDK
Smart Contract	Smart Contract	Smart Contract

Block Chain

Contract Management	Network Management	Alliance Management	Service Management
Peer to Peer Network	Consensus Algorithm	Cryptology Method	Data Storage Technology

| Computing Node 1 | Computing Node 2 | ...... | Computing Node n |

Fig. 7.1 Blockchain's mode of operation

smart contracts between users of the network. Hyperledger and EnergyWeb are good examples of successful platforms that utilize the PoS/PoA consensus algorithm for the power industry (O'Donovan and O'Sullivan 2019).

Despite being originally used for deploying digital cryptocurrency, blockchain technology also has the potential of being a game-changer in the energy sector. According to a 2019 Global Market Insights report, blockchain technology in the energy market is expected to hit USD 18 billion by 2025 (Ankit Gupta 2019). An important incentive that motivates participants in the energy market to adopt blockchain technology would be cost savings. By eliminating the need for intermediaries, complex transaction processes could be simplified and the transaction costs of executing smart energy contracts would decrease (Khatoon et al. 2019). For example, a simulation conducted on a P2P energy transaction platform showed that the P2P transaction unit price was simulated to be 10–30% lower than the fixed unit price determined by a central utility company (Park et al. 2018). Other benefits of blockchain also include enhanced security, heightened transparency, increased efficiency, and most importantly, sustainable decarbonization (Khatoon et al. 2019). In the following sections, we discuss significant use cases of how blockchain technology is being deployed in the energy market, industry and manufacturing sector, as well as its potential in carbon trading.

7.2 Blockchain in the Energy Market

Even before blockchain was adopted by the mainstream, the energy market was already on the path toward significant transformation. Global energy markets are transitioning towards a system characterized by the three key principles: *decarbonization*, *decentralization* and *digitalization*. Yet, the conventional energy market is ill-suited to realize this important transition. Alongside rising energy costs and pressure on environmental sustainability, there is a need to introduce new technologies that could offer cost-saving solutions while aligning with the three key principles as mentioned above. Blockchain could possibly be the promised technology, as reports have shown that it is capable of scaling down costs and increasing efficiency and transparency (Prokop and Koeppen 2018).

The use of blockchain can be categorized into the following groups according to their purposes (Andoni, et al. 2018): (1) decentralized energy trading, (2) cryptocurrencies and investment, (3) IoT, smart devices and automation, (4) smart metering and billing, (5) grid management, (6) electrical e-mobility, and (7) green certificates. As seen from Fig. 7.2, the most popular usage of blockchain in the energy sector is decentralized energy trading, which includes both retail and P2P energy trading.

In the conventional sense, power flows and information flows operate in a unidirectional manner within the electrical power system. Large utilities provide power to the consumer, while information collected from sensors at the consumer's end is transferred back to the utilities for optimal energy dispatch strategies.

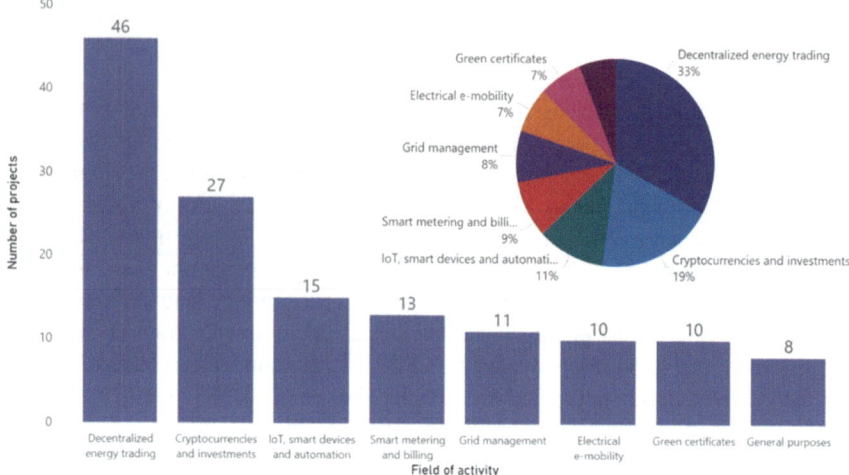

Fig. 7.2 Categorization of blockchain uses in the energy sector from a study of 140 blockchain initiatives (Andoni, et al. 2018)

However, due to the injection of modern technologies such as microgrids, Internet-of-Things (IoT), RES, and intelligent storage, we are now seeing an accelerated paradigm shift towards a bidirectional power and information flow. With this change, previous consumers can now take on a more proactive role by becoming a producer of electrical energy (prosumers) and sell the excess energy generated back to the grid via feed-in-tariffs. However, remuneration levels from feed-in-tariffs are significantly lower than the market electricity price (Immonen et al. 2020). Coupled with decreasing costs and increasing ease of acquiring renewable installations, especially so with the prices of rooftop PV panels, dissatisfied prosumers are now seeking alternatives to P2P energy trading without the need for a central intermediary (Immonen et al. 2020). The increased penetration of these prosumers armed with DERs into the energy market decentralizes the conventional system and renders it obsolete. As the energy market approaches grid parity, it seems reasonable that blockchain would be the ideal solution to satisfy the growing demand for a decentralized energy system.

The launch of Ethereum in mid-2015 allowed the use of smart contracts within blockchain technology, and this significant milestone in blockchain's development was termed Blockchain 2.0 (Dütsch and Steinecke 2017). Smart contracts facilitate automated execution of code within the blockchain that triggers an exchange of value whenever certain conditions are satisfied. The use of blockchain enabled smart contracts is highly useful in P2P energy trading, and the basic concept could be seen in Fig. 7.3. One example would be a blockchain driven P2P energy trading in Brooklyn where 50 homes and businesses connected on the microgrid could buy/sell solar energy among themselves using blockchain-driven smart contracts (Mengelkamp et al. Jan. 2018). The smart meters deployed by the Brooklyn MicroGrid project could measure the energy surplus obtained and convert it into energy tokens. These tokens

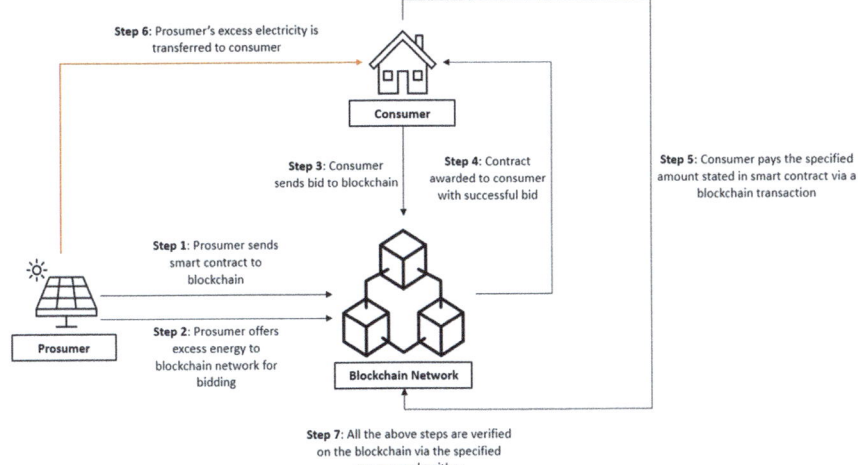

Fig. 7.3 Blockchain concept in P2P energy trading

were then actively traded on the marketplace and transferred from the prosumer's smart wallet to the consumer's smart meter. All these transactions are fully transparent as every member of the community can access the ledger and monitor all transactions personally. Akin to the stock market, the smart contract within the blockchain ledger operates on a bid/offer system. Sellers would present the offer price, while buyers would present the bid price. This two-way price quotation would then be automated by the smart contract when the bid price matches the offer price.

However, the critical issue revolving around P2P energy trading, would be to balance the demand and supply. Due to the varying intermittency of RES and load balancing, blockchain technology alone would not be sufficient to solve this issue. Hence, there is a need to use other intelligent services such as machine learning/artificial intelligence (AI) based prediction, to allow for demand-side flexibility (Andoni, et al. 2018).

Another popular usage of blockchain in the energy sector is cryptocurrencies and investments. Cryptocurrencies in the energy market are a good way of rewarding and incentivizing users to follow desired behaviors and usher in more investments in renewables. For example, cryptocurrency initiatives such as SolarCoin, rewards users with 1 SolarCoin for every 1 MWh of solar energy produced (SolarCoin.org 2020). At its peak in January 2018, each SolarCoin was worth approximately USD$2, and as of September 2020, SolarCoins have been granted to a total of 11 TWh of energy production from solar installations (CoinMarketCap 2020). According to carbon footprint conversion data, comparing the CO_2 emissions of energy obtained from solar to energy obtained from coal, that would equate to a reduction of approximately 9,000 metric megaton of CO_2 equivalent emissions since 2010 (Allen 2011). Therefore, this incentive-based initiative is designed to accelerate the renewable energy transition and lower carbon emission levels.

7.3 Blockchain in the Chemical and Manufacturing Industry

The ongoing fourth industrial revolution (Industry 4.0) emphasizes the automation of industrial practices, particularly in the chemical and manufacturing industries. The main design principles behind Industry 4.0 are transparency, interconnectivity and decentralized operations (United Nations Industrial Development Organization, "Industry 4.0. Opportunities and Challenges of the New Industrial Revolution for Developing Countries and Economies in Transition" 2016). From sensors to cloud computing, Industry 4.0 application requires a large amount of data collection at every node of operation along the supply chain in the manufacturing industry (see Fig. 7.4). Therefore, the timely adoption of blockchain technology alongside Industry 4.0 shows high potential and could prove to be the needed game-changer. In this section, we will look at how blockchain technology could be applied in supply chain management, as well as machine to machine (M2M) communications.

The recent COVID-19 pandemic has revealed to us deep-rooted problems within global chemical/manufacturing supply chains. From the sourcing of raw materials between suppliers and manufacturers, to the global distribution of products between distributors and end-consumers, the entire supply chain has been devastatingly shocked by the impact of COVID-19 (Craven 2020). In fact, PwC's Covid-19 CFO Pulse Survey indicated that supply chain disruptions have been one of the top concerns during this difficult period ("PwC US CFO Pulse Survey : PwC". 2020). Evidently, businesses around the world have learned a painful lesson from this pandemic: there is an urgent need for visibility and digitalization within the supply chain, and the World Economic Forum has proposed that this change could be brought forward with the use of blockchain technology (Liao and Fan 2020).

In an IBM study, approximately 85% of chief supply chain officers feedbacked that managing visibility is one of the pressing issues that they are facing in supply chains (Butner 2010). Any unpredicted disruptions would put to halt the flow of goods

Fig. 7.4 Industry 4.0 application along supply chain of manufacturing industry

along the supply chain, thus damaging the entire business ecosystem. Therefore, higher visibility across supply chains is essential, as it grants real-time tracking of products as they move from one end of the supply chain to the other. In case of any unforeseen disruptions, this enhanced visibility would allow for quick responses from all parties across the supply chain network.

Blockchain offers enhanced visibility via the incorporation of smart contracts (see Fig. 7.5). For example, the supplier prepares the chemical products for shipment, and the shipping receipt is added to the blockchain. While the products are being prepared for vehicular transfer, suppliers/manufacturers would both submit electronic approvals, and this would trigger the smart contract within the blockchain, thus automatically releasing the shipment. Once the shipment arrives at the recipient, he/she would sign another electronic approval, hence authenticating the exchange and transfer of ownership. One example would be TradeLens, a collaboration between IBM and Maersk that uses blockchain to track containers during shipping processes (Dieterich et al. 2020). This way, all parties would have transparency, traceability, and accountability across the supply chain.

Also, trade and payment processes in the chemical and manufacturing industry are still heavily dependent on conventional methods using either spreadsheets or paper-based records. Such conventional methods have proven themselves to be inefficient, error-prone, and costly, due to the inconsistency in non-standardized data entry and out-of-date information transfer between different parties across the blockchain. As seen from government lockdowns and stay-at-home measures being implemented globally due to COVID-19, it is difficult to carry out conventional physical processes. Hence, any sudden disruptions to normality could prove to be damaging and this highlights the importance of digitalizing records within the supply chain.

So, what is the reason behind Blockchain applications within conventional supply chains and how does it accelerate decarbonization? Blockchain provides a secure way

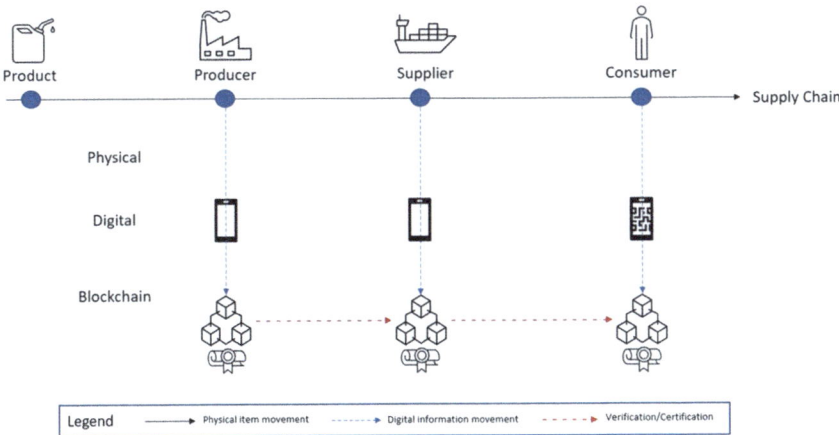

Fig. 7.5 Blockchain-enabled supply chain

for data to be stored on a shared ledger. Due to the immutable nature of blockchain, these data would be verifiable by all parties along the supply chain, while being tamper-proof. At the same time, sensitive data could be made private, as blockchain operates via consensus algorithms, thus allowing users to capture the proof of the data, and not the data itself. Hence, digitalization of records using blockchain promotes trust between stakeholders while being efficient and cost-saving.

Blockchain also has a direct positive influence on environmentally sustainable supply chains. Improved visibility and digitalization of records, would help optimize the flow of goods and information across supply chains. Consequentially, it reduces the need for unnecessary recalling/reworking of goods. An efficient vehicular dispatch strategy in place for the transportation of goods across supply chains, decreases excessive consumption and reduces greenhouse gas (GHG) emissions in the process (Saberi et al. Apr. 2019). Blockchain transparency also empowers consumers with product knowledge on how their purchased products were manufactured and handled across the complex supply chain. This is important considering consumers nowadays are environmentally aware, yet are unable to translate these concerns into green purchases (William et al. 2010). To help this cause, an ongoing project named "Provenance" has successfully implemented a digital platform for consumers to trace the movement of their products along the supply chain (Wood et al. 2015).

Blockchain can also be applied in M2M communications within Eco-Industrial Parks (EIPs). EIPs comprise of integrated infrastructures that promote synergy and product integration between chemical companies and utilities. EIPs have been popular due to the many benefits attained via the symbiosis, such as shared resources and reduction in waste and pollution (Zhou et al. 2017). As seen in Singapore, Jurong Island is an artificial island that has a portfolio of over 100 chemical plants, and the existing shared infrastructure boasts shared processes such as steam networks, wastewater treatment flows and electricity generation. For example, Sembcorp Industries, a utilities company, adopts a circular economy approach and treats industrial wastewater from other chemical plants on Jurong Island before re-supplying it back to them for industrial processes. The benefits of scale from such business practices justifies the purpose of EIPs. However, EIPs account for a significant amount of greenhouse gas emissions (Pan et al. Aug. 2015). To reduce these emissions, the increased efficiency within these chemical industrial parks should be prioritized, and this could be achieved through M2M communication.

M2M communication is the bedrock of most IoT applications. Blockchain, being an append-only public ledger, allows IoT applications to function with high transparency. Furthermore, smart contracts could seamlessly integrate automated transactions into the blockchain, hence eliminating trust issues and third-party involvement in any transaction. For example, when smart sensors get alerted that a certain chemical reactant in the tank falls below a predefined level, the smart contract acts as a guarantee, and triggers an automated order of the reactant directly from the buyer to the vendor. Hence there is no longer a need to carry out such transactions via intermediaries such as banks and credit card companies, thus resulting in lower transaction fees and higher efficiency. Other uses of blockchain on the IoT platform include monitoring of equipment, predictive maintenance, and process optimization (Higgins and Sandner 2020).

7.4 Blockchain and Carbon Trading

In the 1997 Kyoto protocol, market mechanisms for carbon trading were introduced to tackle climate change by reducing the amount of global GHG emissions (Hepburn Nov. 2007). One of the market mechanisms, the emissions trading scheme (ETS), works on a cap-and-trade market. Different industrial sectors are allocated a limited number of permits that allows them to produce a certain amount of GHG. Entities left with excess unused permits can sell them to other entities that have exceeded their GHG production limit. Therefore, this zero-sum game allows for a clear GHG emission target to be achieved.

Historically, it has proven to be effective in sustainable decarbonization. A study across 6 major regions has shown that 10 years of ETS implementation has contributed to approximately 23.4% of CO_2 reduction (Villoria-Sáez et al. 2016). In the European Union (EU) alone, ETS has saved 1.2 billion tons of CO_2 from 2008 to 2016, which is a 3.8% reduction compared to a non-existing carbon market (Bayer and Aklin 2020).

However, the low carbon prices and uncertainty in the future of the carbon market have undermined the effectiveness of ETS and discouraged significant investments in the long run (Mo et al. 2015). There are also other complications within the carbon markets such as fraud and risk of double counting (Dufrasne 2020). Therefore, there is a need to introduce new technologies that can increase the effectiveness of the carbon trading market.

There are several advantages of integrating blockchain into the current ETS-backed carbon market. Blockchain provides high-level transparency while ensuring the accuracy and privacy of information due to its immutable nature (Pan et al. 2019). Every single transaction being made on the carbon market would have its own unique timestamp and cryptographic hash key, thus ensuring historical trace-back. This helps in forced accountability, and participants in the carbon market must be responsible for themselves. In turn, the real-time monitoring of carbon credits transfer across the carbon market reduces the risk of fraud and double counting (Schneider et al. Aug. 2015).

Another proposed method of blockchain integration into the carbon market is to incorporate a reputation-based trading system into the conventional ETS model (Khaqqi et al. Jan. 2018). Each entity in the carbon market would be allocated a reputation index, which is a function of past emission rates and emissions reduction strategy. In Article 12 of the Kyoto Protocol, the Clean Development Mechanism (CDM) was designed to encourage developed countries and businesses to invest in emission-reduction projects in either their own country or other developing countries. Entities with lower GHG emissions per product ratio and higher additionality in CDM projects (M. M. Lazarus Carrie Lee Pete Erickson Randall Spalding-Fecher

2016), would be given a higher reputation index. Consequentially, the higher the entities' reputation, the wider the access given to them. Such a reputation-based system provides financial incentives, granting these entities more opportunities to select better bids/offers. Ultimately, the end goal would be to encourage more participants in the carbon market to adopt long-term solutions to reducing GHG emissions (Khaqqi et al. Jan. 2018).

7.5 Outlook

Overall, blockchain applications in energy, manufacturing and carbon trading systems would prove beneficial to conventional industrial practices and reduce carbon emissions as a positive spillover effect. For example, IBM and Energy Blockchain Labs Inc. created a blockchain platform that tracks the carbon footprint of activities along the supply chain. Their goal is to have a 20–50% reduction of the 10-months average carbon asset lifecycle, and this would channel more investments into green initiatives and reduce carbon emissions in the long run ("Energy Blockchain Labs Inc. I IBM", IBM, Energy blockchain labs inc. 2018). The previously discussed SolarCoin and other blockchain-based reward systems for renewable energy producers could further decarbonize the energy sector. We estimate blockchain technologies could bring more than a 30% reduction in both emissions and cost in the next decade, with the improved system efficiency and increase adoption of clean technologies.

References

Allen S (2011) Carbon footprint of electricity generation, Jun 2011. www.parliament.uk/post. Accessed 10 Sep 2020

Andoni M et al (2019) Blockchain technology in the energy sector: a systematic review of challenges and opportunities. Renew Sustain Energy Rev 100:143–174. https://doi.org/10.1016/j.rser.2018. 10.014

Ankit Gupta ASB (2019) Blockchain In energy market trends. Global Industry 2019–2025 Report, Apr 2019. https://www.gminsights.com/industry-analysis/blockchain-in-energy-market?utm_source=prnewswire.com&utm_medium=referral&utm_campaign=Paid_prnewswire. Accessed 04 Aug 2020

Bayer P, Aklin M (2020) The European Union emissions trading system reduced CO_2 emissions despite low prices. Proc Natl Acad Sci U S A 117(16):8804–8812. https://doi.org/10.1073/pnas. 1918128117

Butner K (2010) The smarter supply chain of the future. Strategy Leadership 38(1):22–31. https://doi.org/10.1108/10878571011009859

CoinMarketCap (2020) SolarCoin (SLR) price, charts, market cap, and other metrics. CoinMarketCap, Sep 2020. https://coinmarketcap.com/currencies/solarcoin/. Accessed 10 Sept 2020

Craven M (2020) Coronavirus' business impact: evolving perspective. McKinsey. https://www.mckinsey.com/business-functions/risk/our-insights/covid-19-implications-for-business. Accessed 8 Aug 2020

Dieterich V, Ivanovic M, Meier T, Zäpfel S, Utz M, Sandner P (2020) FSBC Working paper application of blockchain technology in the manufacturing industry. http://explore-ip.com/2017_Blockchain-Technology-in-Manufacturing.pdf. Accessed 8 Aug 2020

Dufrasne G (2020) Carbon markets 101 THE ULTIMATE GUIDE TO GLOBAL OFFSETTING MECHANISMS. https://carbonmarketwatch.org/wp/wp-content/uploads/2019/06/CMW-CARBON-MARKETS-101-THE-ULTIMATE-GUIDE-TO-MARKET-BASED-CLIMATE-MECHANISMS-WEB-FINAL-SINGLE.pdf. Accessed 12 Aug 2020

Dütsch G, Steinecke N (2017) Use cases for Blockchain Technology in energy & commodity trading. www.pwc.com. Accessed 31 Jul 2020

Energy Blockchain Labs Inc (2018) IBM, Energy Blockchain Labs Inc. https://www.ibm.com/case-studies/energy-blockchain-labs-inc. Accessed 27 Aug 2020

FSU Centre (2003) Global trends in renewable energy investment 2020. https://www.fs-unep-centre.org/wp-content/uploads/2020/06/GTR_2020.pdf. Accessed 04 Aug 2020

Hepburn C (2007) Carbon trading: a review of the kyoto mechanisms. Annu Rev Environ Resour 32:375–393. https://doi.org/10.1146/annurev.energy.32.053006.141203

Higgins M, Sandner P (2020) Blockchain business models for autonomous IoT sensor devices. http://explore-ip.com/2019_Blockchain-Business-Models-for-Autonomous-IoT-Sensor-Devices.pdf. Accessed 8 Aug 2020

Immonen A, Kiljander J, Aro M (2020) Consumer viewpoint on a new kind of energy market. Electr Power Syst Res 180. https://doi.org/10.1016/j.epsr.2019.106153

Khaqqi KN, Sikorski JJ, Hadinoto K, Kraft M (2018) Incorporating seller/buyer reputation-based system in blockchain-enabled emission trading application. Appl Energy 209:8–19. https://doi.org/10.1016/j.apenergy.2017.10.070

Khatoon A, Verma P, Southernwood J, Massey B, Corcoran P (2019) Blockchain in energy efficiency: potential applications and benefits. Energies 12(17):1–14. https://doi.org/10.3390/en12173317

Lazarus M, Lee CM, Erickson P, Spalding-Fecher R (2016) How additional is the clean development mechanism? Analysis of the application of current tools and proposed alternatives. www.oeko.de. Accessed 12 Aug 2020

Liao R, Fan Z (2020) Supply chains are in chaos. Here's how to make them more resilient. World Economic Forum, 6 Apr 2020. https://www.weforum.org/agenda/2020/04/supply-chains-resilient-covid-19/. Accessed 12 Aug 2020

Mengelkamp E, Gärttner J, Rock K, Kessler S, Orsini L, Weinhardt C (2018) Designing microgrid energy markets: a case study: the Brooklyn Microgrid. Appl Energy 210:870–880. https://doi.org/10.1016/j.apenergy.2017.06.054

Mo J-L, Agnolucci P, Jiang M-R, Fan Y (2015) The impact of Chinese carbon emission trading scheme (ETS) on low carbon energy (LCE) investment. Energy Policy 89:271–283. https://doi.org/10.1016/j.enpol.2015.12.002

O'Donovan P, O'Sullivan DTJ (2019) A systematic analysis of real-world energy blockchain initiatives. Futur Internet 11(8). https://doi.org/10.3390/fi11080174

Pan M et al (2015) Applying industry 4.0 to the Jurong Island Eco-industrial park. Energy Procedia 75:1536–1541. https://doi.org/10.1016/j.egypro.2015.07.313

Pan Y et al (2019) Application of blockchain in carbon trading. Energy Procedia 158:4286–4291. https://doi.org/10.1016/j.egypro.2019.01.509

Park LW, Lee S, Chang H (2018) A sustainable home energy prosumer-chain methodology with energy tags over the blockchain. Sustain 10(3):1–19. https://doi.org/10.3390/su10030658

Prokop M, Koeppen M (2018) Blockchain: a true disruptor for the energy industry use cases and strategic questions. https://www2.deloitte.com/content/dam/Deloitte/us/Documents/energy-resources/us-blockchain-disruptor-for-energy-industry.pdf. Accessed 04 Aug 2020

PwC US CFO Pulse Survey: PwC (2020). https://www.pwc.com/us/en/library/covid-19/pwc-covid-19-cfo-pulse-survey.html. Accessed Aug 8 2020

Saberi S, Kouhizadeh M, Sarkis J, Shen L (2019) Blockchain technology and its relationships to sustainable supply chain management. Int J Prod Res 57(7):2117–2135. https://doi.org/10.1080/00207543.2018.1533261

Schneider L, Kollmuss A, Lazarus M (2015) Addressing the risk of double counting emission reductions under the UNFCCC. Clim Change 131(4):473–486. https://doi.org/10.1007/s10584-015-1398-y

SolarCoin.org (2020) One megawatt hour one SolarCoin. https://solarcoin.org/wp-content/uploads/SolarCoinPresentation.pdf. Accessed 4 Aug 2020

United Nations Industrial Development Organization (2016) Industry 4.0. Opportunities and challenges of the new industrial revolution for developing countries and economies in transition. 2030 Agenda Sustain Dev Goals. https://doi.org/10.1007/978-1-4842-2047-4

Villoria-Sáez P, Tam VWY, del Río Merino M, Arrebola CV, Wang X (2016) Effectiveness of greenhouse-gas emission trading schemes implementation: a review on legislations. J Clean Prod 127:49–58. https://doi.org/10.1016/j.jclepro.2016.03.148

William Y, Kumju H, Seonaidh M, Caroline (2010) Sustainable consumption: green consumer behaviour when purchasing products. Sustain Dev J 18(1):20–31. https://doi.org/10.1002/sd.394

Wood G, Meiklejohn S, Buchanan A, Brewster C (2015) Blockchain: the solution for supply chain transparency. Provenance, 21 Nov 2015. https://www.provenance.org/whitepaper. Accessed 20 Aug 2020

Zhou L et al (2017) Towards an ontological infrastructure for chemical process simulation and optimization in the context of eco-industrial parks. Appl Energy 204:1284–1298. https://doi.org/10.1016/j.apenergy.2017.05.002

Part III
Sectors & Impact

Chapter 8
Cyber Physical Production Systems and Their Role for Decarbonization of Industry

Sebastian Thiede

Abstract Industry is the economic sector with the highest contribution to global Greenhouse Gas Emissions (GHG) and therewith plays a major for future decarbonization. The question arises whether cyber physical production systems (CPPS) as one core element of the digital transformation of industry can contribute here. To derive the most promising fields of action and investigate the role of CPPS a holistic perspective on the industry sector is necessary. Besides energy efficiency also fostering energy transition towards renewable sources as well as material efficiency turn out to be important leverages. Major CPPS based contributions can be expected through innovative, advanced control approaches—especially for complex production situations with changing products and diverse influencing factors. But also in other areas at least indirect contributions through CPPS can be expected, e.g. for identification of best practice technologies, to align energy demand and renewable energy supply or to support material efficiency-related improvements. Altogether, CPPS based potential for industry is estimated in a range of 15–25%.

8.1 Introduction

Industry plays a crucial role in today's societies and has very strong relevance from an economic, environmental and social perspective. From the environmental side, industry is for example of very strong relevance in the context of global greenhouse gas emissions (GHG). According to the IPCC, industry is responsible for 21% of direct emissions (largely through the burning of fossil fuels, e.g. for process heating, and also process emissions) but also—through its electricity and steam demand covered by external power plants—for further 11% of indirect emissions (IPCC 2014) (Fig. 8.1). Therewith, with a share of almost a third of the total GHG emissions, industry is the economic sector with the highest contribution and therewith a major field of action for decarbonization.

S. Thiede (✉)
Chair of Manufacturing Systems, Faculty of Engineering Technology, Department of Design, Production & Management, University of Twente, 7522LW Enschede, The Netherlands
e-mail: s.thiede@utwente.nl

© Springer Nature Switzerland AG 2022
O. Inderwildi and M. Kraft (eds.), *Intelligent Decarbonisation*, Lecture Notes in Energy 86, https://doi.org/10.1007/978-3-030-86215-2_8

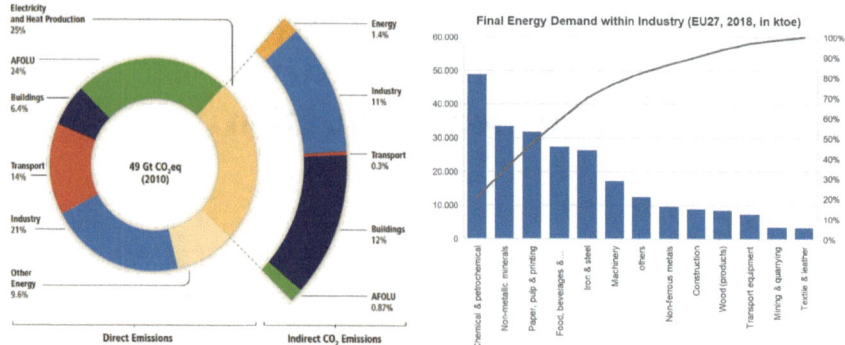

Fig. 8.1 Greenhouse gas emissions by economic sectors (IPCC 2014) and energy use of different industry branches (Eurostat 2020)

The industry sector covers different branches which can be broadly differentiated according to the types of processes and resulting products. Process industry is depicted by more continuous processes often for producing large quantities of materials (e.g. chemical, paper, steel) which are typically an input for discrete manufacturing with a focus on defined volumes of individual pieces (e.g. machinery, transportation equipment). Figure 8.1 also gives an overview of the energy demand of the different branches. According to that, the most relevant branches are chemical, iron and steel, building materials, paper and food industry (all rather process industry) followed by machinery and transportation equipment (Eurostat 2020).

Industry is a sector of continuous and dynamic changes. Besides the rising pressure in the context of necessary climate change actions, it is currently under the strong influence of ongoing digitalization (associated with terms like Industry 4.0/Smart Manufacturing) with impacts in manifold ways (Kagermann et al. 2013; Kang et al. 2016). While many technologies, methods and terms are associated with this 4th industrial revolution, cyber physical production systems (CPPS) can actually be seen as one core element of those developments (Monostori et al. 2016). CPPS (Fig. 8.2)

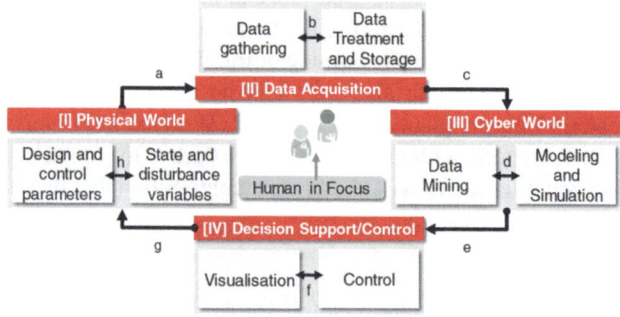

Fig. 8.2 Cyber physical production system framework (Thiede 2018)

stand for the connection of the "physical world" in the sense of the considered factory systems (or a specific element within that) and a "cyber" subsystem that describes an up-to-date virtual representation based on different possible modelling approaches (e.g. data-based models, physical models, geometrical representations).[1] The cyber world is updated through appropriate data acquisition (e.g. process/machine parameters and state variables) and allows advanced analytics or forecasting in order to facilitate an improved operation of manufacturing. The loop is closed through the feedback and actual intervention in the physical world through decision support for involved stakeholders or even direct/automatic control (Thiede 2018).

Different solutions based on CPPS logic can be found in research and practice which underline the general potential to improve production in terms of different objectives. Against this background, this chapter addresses the question of whether and to what extent CPPS can support the decarbonization of the manufacturing industry as well.

8.2 Energy Flows in Industry and CPPS

As a base for further analysis of the impact of CPPS in the context of decarbonization, it is important to understand the system boundaries, the structure of industry as well as the related energy flows.

8.2.1 System Definition

Within industry four different types of manufacturing entities can be distinguished which typically build up a connected value chain: extractive industry (e.g. mining of iron ore) delivers raw resources to the materials industry (e.g. steel, aluminium, chemicals) which converts those into raw materials that can be used in manufacturing and construction (e.g. transportation equipment) in order to produce physical goods for customers (Fig. 8.3, IPCC 2014). Waste from industry and market is handled through the waste industry which to some extent brings back remanufacturing products/parts or recycled materials into the value chain. At all stages, very significant quantities of energy are needed which leads to respective energy-related emissions (plus some direct process emissions e.g. through chemical reactions)—as indicated above, materials industry is clearly leading here. This underlines the important interrelation of material and energy demand (e.g. with the embodied energy of material as an important indicator) in the context of their role toward decarbonization—saving material can avoid energy related emissions in upstream industry.

---

[1] In research and practice also terms like e.g. Digital Twin, Digital Shadow, Virtual Twin can be found which—sometimes with slightly different emphasis on certain elements—represent similar concepts.

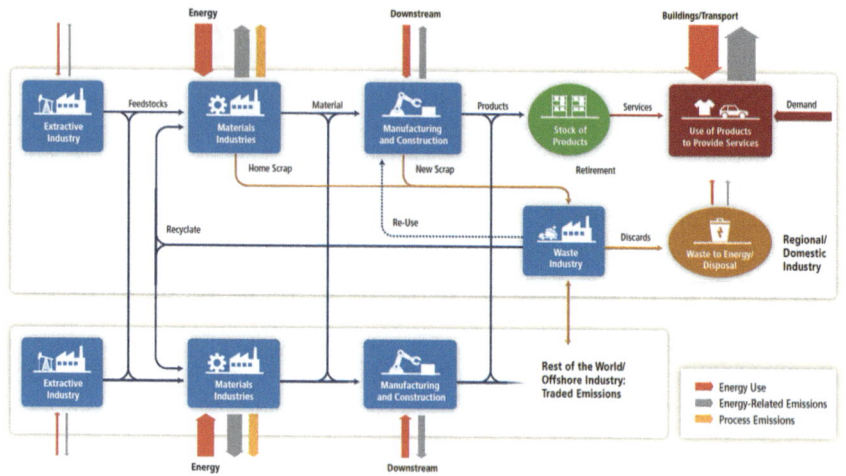

Fig. 8.3 Industry sector structure and interactions (IPCC 2014)

Industrial activities typically take place in dedicated facilities (often referred to as factories) that consist of three different subsystems: value adding (production) equipment, technical building services (TBS, e.g. for internal energy conversion and providing necessary production conditions like environmental temperature, moisture or purity) and the building shell. Those subsystems are connected through internal energy, material and information flows (Herrmann et al. 2014). Material flows consist of raw materials, auxiliary materials, waste streams and (semi)finished products which are the main purpose of manufacturing. In terms of energy different energy carriers (e.g. electricity, gas, oil, compressed air) play a role. They are externally acquired or internally generated or converted. Information flows allow the status monitoring of operations and therewith planning and control (Fig. 8.4).

8.2.2 Industrial Energy Flows and Related Emissions

Over the industrial value chains energy demand is ultimately the main cause of industrial emissions. According to the system definition given above, process and nonprocess energy can be distinguished. Process energy is the amount of energy that is directly used in the value adding process, typically in form process heating and cooling, electro-chemical or mechanical energy (e.g. electrical drives) (US Department of Energy 2019). Nonprocess energy is needed for supporting industrial activities and often related to the technical building services of factories, e.g. space heating/cooling or lighting play an important role here. Fig. 8.5 shows a Sankey diagram of energy flows in the US manufacturing sector. It underlines that over all sectors process energy is dominating and sum up to a share of approx. 88% of energy

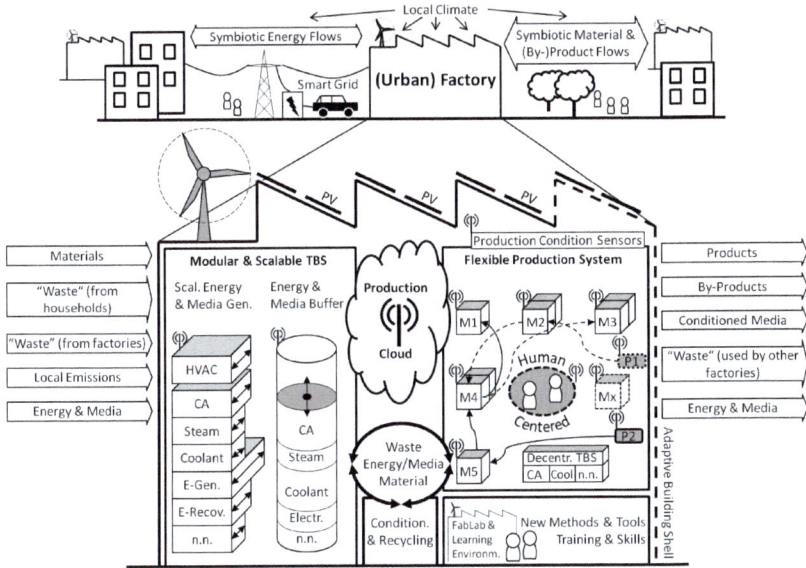

Fig. 8.4 Holistic understanding for factories of the future (Herrmann et al. 2014)

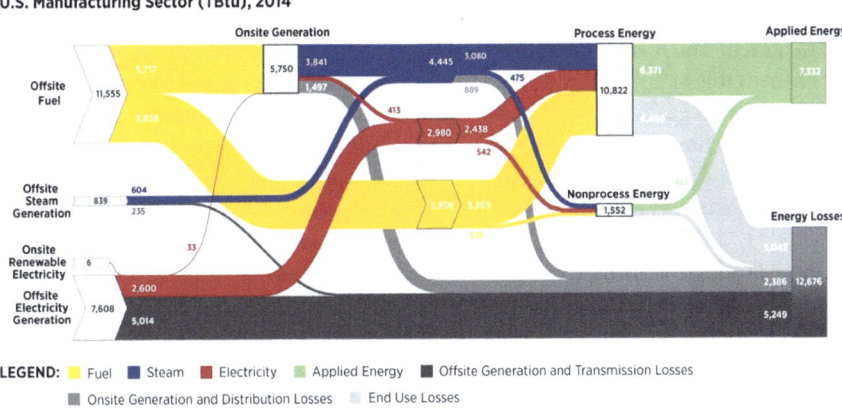

Fig. 8.5 Energy flows in US manufacturing sector (US Department of Energy 2019)

demand (US Department of Energy 2019). However, this ratio very much depends on the specific industry sector and even specific factory, e.g. for discrete manufacturing the share of nonprocess energy is typically significantly higher. Even more, this is an energy balance (in Btu) which naturally puts an emphasis on fuel/heat-related flows. In absolute energy values, electricity seems to have less relevance for manufacturing but obviously this leaves out the inefficiencies of the energy generation chain and therewith the actual impact on related carbon emissions.

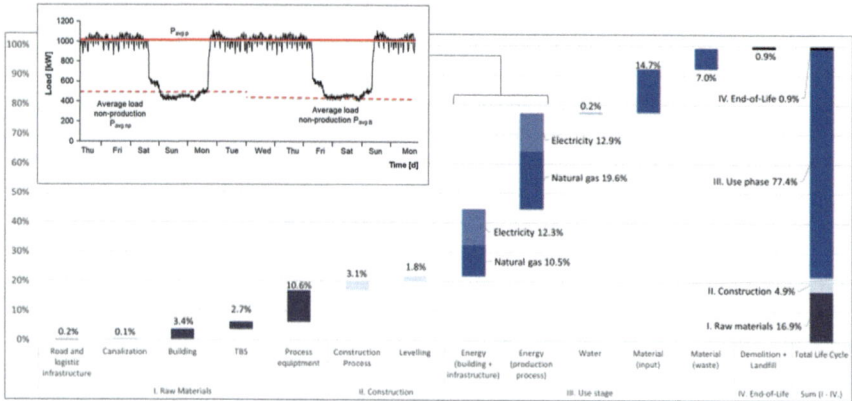

Fig. 8.6 Greenhouse gas emissions of an automotive factory over life cycle (Gebler et al. 2020), exemplary load profile from Dehning et al. (2019)

Therefore, Fig. 8.6 exemplary shows the results of a Life Cycle Assessment of an automotive factory with a focus on Global Warming Potential (in CO_2, eq) as the usual impact category for GHG (Gebler et al. 2020). This analysis focuses on the factory itself (over a time of 30 years) as a system boundary so raw materials for the product (car) are not included here while their relation emissions are not caused by the factory itself. However, auxiliary materials that are needed for the processes are included. The results underline that the energy demand is dominating the carbon footprint of this factory. Over 30% of the total emissions are caused by energy demand (electricity and gas) of the production processes itself, further approx. 22% through the related technical building services. Besides this aggregated perspective on energy demand also the time-related aspects need to be taken into account, a typical load curve of a factory is also shown in Fig. 8.6. It underlines that the energy base load of a factory is a very relevant aspect that needs to be addressed—even without value-creating production around 50–60% of the energy is used due to active TBS processes and high idle energy demand of production machines (Dehning et al. 2019).

8.2.3 Leverages for Decarbonization Through CPPS

Based on the elaborations in the last sections, three quite interrelated main fields of action for addressing decarbonization in industry can be identified (Fig. 8.7) (IPCC 2014).[2] First, energy efficiency stands for the improvement of the ratio of product output and the necessary energy input, e.g. through the selection of energy-efficient

---

[2] The focus here is on industry-related activities with the potential impact of process-oriented CPPS. Of course there are further, more product and customer-oriented fields of action like material-efficient product design, sustainable consumption, sharing concepts and circular economy [IPCC 2014].

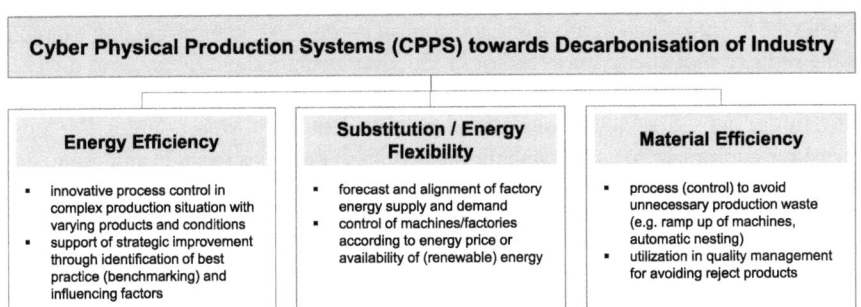

Fig. 8.7 Overview of potential CPPS application fields for industrial decarbonization

technologies or better process/factory control. Secondly, substitution (or effectiveness) strategy aims at changing the source of energy generation while switching to renewable resources. Thirdly, material efficiency is also a strategy—in the manufacturing domain. This is related to avoiding waste over the value chain through better quality rates (fewer rejects) and better utilization as well as recycling of material. In general, CPPS can contribute to all those areas based on the acquisition of data and intelligent models that are embedded in decision support or automatic control systems.

Energy Efficiency

For the area of energy efficiency, the selection of the most efficient technologies and the correct dimensions according to the designated working tasks can be seen as major leverage for improvement. Studies underline improvement potentials in the range of 20–25% through applying commercially available best-practice technologies (Fawkes et al. 2016; Chang et al. 2015; IPCC 2014), further potential 20% are estimated through innovations in the future (IPCC 2014). This applies to both direct production equipment but also strongly to the technical building services. However, technology selection and dimensioning are more related to the planning phase where limited continuous data of considered equipment is available. Still, CPPS can make at least an indirect contribution here. Examples are CPPS where e.g. machine learning is used to systematically identify best practices among other production machines or production areas within a company. Other approaches help to analyse the most influencing factors on energy demand (Dehning et al. 2017; Thiede et al. 2020). With that, decision support can be given for the strategic (re)planning of factories. The more direct impact and full potential of CPPS can be utilized when it comes to the operation phase of industrial facilities. Based on up to real-time data streams and intelligent models, the parameters of running processes (e.g. temperature setting, machine speeds) can be ideally adapted according to the current production situation and the requirements of the product. Studies mention saving potentials over all industry sectors of 5–15% in the area of process individual but specifically for more integrated control approaches (Chang et al. 2015). Of course, those are not necessarily CPPS based control schemes but CPPS lower the entrance barrier and enable the development of more holistic control models. Another promising example is the reduction of energy baseload of production machines and factories as a whole (Dehning et al. 2019). CPPS can support here by analysing the current and short-term future state of production and deriving intelligent shutdown strategies. For those measures equipment of technical building services (TBS) is a feasible and promising application domain. As previous sections showed, their "nonprocess" energy can be a very relevant leverage for improvement and TBS strongly contribute to the baseload. Due to their technical setting (e.g. often over dimensioned, flexible control possible) and the just indirect coupling to the actual value-adding process chain, TBS typically also allow some more degree of freedom for changes.

Substitution/Energy Flexibility

The substitution to renewable energy sources is a very powerful strategy that can significantly reduce energy-related carbon emissions. The obvious strategy for industrial companies is to just acquire renewable-based energy from the energy supplier which is a matter of pricing and availability in the end. However, over the last years also more proactive strategies have become more relevant for industry while companies produce their own renewable-based energy and/or become an active part of the energy market. Besides potentially providing energy from own (renewable) sources to the market, also the concept of energy flexibility is of increasing interest. Within that, companies dynamically change their energy demand behaviour over time depending on e.g. the market price and/or the availability of (either onsite or grid) renewable energy (Sauer et al. 2019). Thus, it can be beneficial from both the environmental but also the cost side. Energy flexibility can play an important role in the energy transition of industry and society as a whole. CPPS can support at this point through forecasting and aligning energy supply and demand (e.g. through control of machines), ideally without harming the production output. Respective approaches can be found in research (e.g. Beier et al. 2017, Sauer et al. 2019).

Material Efficiency

Due to the energy intensity of the materials industry and the later inefficient usage of those materials in manufacturing, material efficiency is a relevant field of action for industrial decarbonisation as well. Improvement potentials can be found in avoiding waste materials or quality rejects as well as fostering recycling concepts. Studies have shown that companies estimate potential material savings with 5–10% (Schröter et al. 2011) through advanced manufacturing processes, new production technologies, circular concepts but also improved product designs. There is also a strong connection to other manufacturing fields of action like quality management. Digitalization is widely seen as an important enabler to seize this potential (Neligan and Schmitz 2017), CPPS based approaches play an interesting role here. For example, CPPS can help to derive root causes of material inefficiencies and low quality rates which allow improved planning and control of processes (e.g. improved ramp-up control, automatic nesting). Furthermore, soft sensor approaches (virtual metering point) can be established which predict the product properties based on current process parameters and state variables. Without too much additional effort for physical measurements, this allows a dense monitoring of quality, avoids further processing of low-quality parts or even enables the adaption of downstream production processes (Filz et al. 2020).

8.2.4 Related Challenges

As shown, CPPS based approaches can make a significant contribution to industrial decarbonization within different fields of action. But of course, also the related challenges need to be discussed. For important improvement measures like technology selection and dimensioning, complex CPPS are not necessarily required while still important steps towards decarbonization can be made. Even more, also for general process improvements within operation temporary analysis might be sufficient to derive better solutions which just need to be implemented once for taking effect. In contrast to that, CPPS based approaches enfold their full potential in case of complex production situations with many influencing factors, changing products and/or production conditions. They are also of specific benefit when just through CPPS implementation continuous impact and improvement can be achieved.

It is important to take into account that CPPS implementation comes with some efforts in terms of costs but also environmental impact (EI). As indicated in Fig. 8.8 new components (e.g. computers, sensors, screens, cables) need to be introduced which carry an additional environmental backpack and also continuously cause additional environmental impact during their operation (Thiede 2018). To avoid unfavourable rebound effects, it is obviously a minimum goal to achieve an environmental break-even thus saving enough energy in the application to compensate those

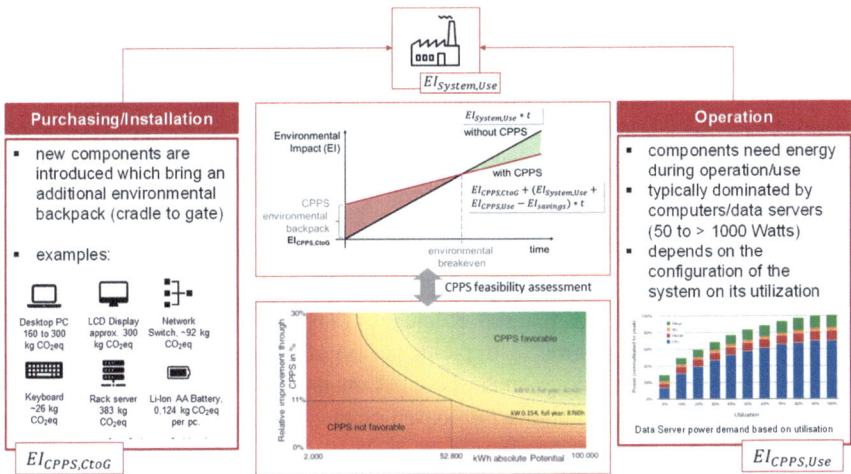

Fig. 8.8 Environmental efforts and feasibility of implementing CPPS in industry (adapted from Thiede 2018, 2021 with data from Barroso et al. 2013; Teehan and Kandlikar 2013, ecoinvent 2020)

additional efforts. Besides appropriate CPPS design and selection of components, a careful definition of use cases that have sufficient improvement potential is needed. Feasibility diagrams can support here to identify favourable and non-favourable CPPS use cases based on the absolute potential (in kWh) that is addressed combined with necessary relative improvement impact over a defined time frame (Fig. 8.8).

8.3   Conclusions and Outlook

For the assessment of decarbonization potentials of cyber physical production system (CPPS) in industry, a holistic perspective is necessary. Considering the different types of industry and their interrelations reveals that besides energy efficiency and energy substitution also material efficiency is a relevant field of action. In general, CPPS can serve as a facilitator for the identification and implementation of many related improvement measures.

CPPS are specifically feasible and can make quite direct contributions when it comes to the operation of manufacturing equipment and factories as a whole. They enfold their full potential in complex production situations with e.g. varying products, a diversity of process parameters and state variables as well as further internal and external influencing factors. Besides consideration of single processes, this also fosters promising CPPS applications on the system level, e.g. for superior control of different factories elements. In this context, special attention should be also paid to technical building services which bear interesting potential for improvement, e.g. for adapting to the current production situation and external conditions (e.g. weather). TBS related measures are also an interesting leverage when it comes to reducing the energy baseload of factories. Existing overarching studies (Fawkes et al. 2016) and also current specific examples (Schulze et al. 2019; Thiede et al. 2020) underline that specifically those integrated control systems offer energy improvement potentials of over 15%. CPPS can play a crucial role here and enable control of more complex systems while also lowering the entrance barriers to seize those potentials (e.g. identification of most important factors, scalable machine learning models that can be operated online). As mentioned before, CPPS can also make valuable at least indirect contributions in the planning phase in order to improve dimensioning and identification of the best available technologies. Further CPPS based potentials can be found through supporting energy flexibility (which is still a bit limited in terms of industrial coverage) and material efficiency. Especially for the latter there are favourable synergies (in terms of implementation efforts and benefits) with other manufacturing related activities such as quality management. Altogether, the industrial decarbonization potential of CPPS based on the current technological level can be estimated at 15–25%.

References

Barroso LA, Clidaras J, Hölzle U (2013) The datacentre as a computer: An introduction to the design of warehouse-scale machines. Synth Lect Comput Arch 8(3):1–154

Beier J, Thiede S, Herrmann C (2017) Energy flexibility of manufacturing systems for variable renewable energy supply integration: real-time control method and simulation. J Clean Prod 141:648–661

Chan Y, Kantamaneni R, Allington M (2015) Study on energy efficiency and energy saving potential in industry and on possible policy mechanisms. ICF Consulting Limited, London. Tratto il giorno, 6(08) (2016)

Dehning P, Thiede S, Mennenga M, Herrmann C (2017) Factors influencing the energy intensity of automotive manufacturing plants. J Clean Prod 142:2305–2314

Dehning P, Blume S, Dér A, Flick D, Herrmann C, Thiede S (2019) Load profile analysis for reducing energy demands of production systems in non-production times. Appl Energy 237:117–130

Eurostat (2020), Energy statistics, Energy Balances, https://ec.europa.eu/eurostat/web/energy/data/energy-balances

Fawkes S, Oung K, Thorpe D (2016) Best practices and case studies for industrial energy efficiency improvement—an introduction for policy makers. UNEP DTU Partnership, Copenhagen

Filz MA, Gellrich S, Turetskyy A, Wessel J, Herrmann C, Thiede S (2020) Virtual quality gates in manufacturing systems: framework, implementation and potentials, accepted for publication. J Manuf Mater Process, Spec Issue: Cyber Phys Prod Syst

Gebler M, Cerdas F, Thiede S, Herrmann C (2020) Life cycle assessment of an automotive factory: Identifying challenges for the decarbonization of automotive production–A case study. J Clean Prod, 122330

Herrmann C, Schmidt C, Kurle D, Blume S, Thiede S (2014) Sustainability in manufacturing and factories of the future. Int J Precis Eng Manuf-Green Technol 1(4):283–292

IPCC (2014) Climate change 2014: mitigation of climate change. Contribution of Working Group III to the Fifth Assessment Report of the Intergovernmental Panel on Climate Change [Edenhofer O, Pichs-Madruga R, Sokona Y, Farahani E, Kadner S, Seyboth K, Adler A, Baum I, Brunner S, Eickemeier P, Kriemann B, Savolainen J, Schlömer S, von Stechow C, Zwickel T, Minx JC (eds)]. Cambridge University Press, Cambridge, United Kingdom and New York, NY, USA

Kagermann H, Helbig J, Hellinger A, Wahlster W (2013) Recommendations for implementing the strategic initiative Industrie 4.0: Securing the future of German manufacturing industry; final report of the Industrie 4.0 Working Group, 2013

Kang HS, Lee JY, Choi S, Kim H, Park JH, Son JY, Kim BH, Do Noh S (2016) Smart manufacturing: Past research, present findings, and future directions. Int J Precis Eng Manuf-Green Technol 3(1):111–128

Monostori L, Kádár B, Bauernhansl T, Kondoh S, Kumara S, Reinhart G, Sauer O, Schuh G, Sihn W, Ueda K (2016) Cyber-physical systems in manufacturing. CIRP Ann 65(2):621–641

Neligan A, Schmitz E (2017) Digitale Strategien für mehr Materialeffizienz in der Industrie: Ergebnisse aus dem IW-Zukunftspanel, IW-Report, No. 3/2017, Institut der deutschen Wirtschaft (IW), Köln

Sauer A, Abele E, Buhl HU (2019) Energieflexibilität in der deutschen Industrie: Ergebnisse aus dem Kopernikus-Projekt-Synchronisierte und energieadaptive Produktionstechnik zur flexiblen Ausrichtung von Industrieprozessen auf eine fluktuierende Energieversorgung (SynErgie). Fraunhofer Verlag

Schröter M, Lerch C, Jäger A (2011) Materialeffizienz in der Produktion: Einsparpotenziale und Verbreitung von Konzepten zur Materialeinsparung im Verarbeitenden Gewerbe. A report prepared for the German Federal Ministry of the Economy and Technology. Karlsruhe, Germany

Schulze C, Thiede S, Thiede B, Kurle D, Blume S, Herrmann C (2019) Cooling tower management in manufacturing companies: a cyber-physical system approach. J Clean Prod 211:428–441

Teehan P, Kandlikar M (2013) Comparing embodied greenhouse gas emissions of modern computing and electronics products. Environ Sci Technol 47(9):3997–4003. ecoinvent database version 3.3, https://www.ecoinvent.org

Thiede S (2018) Environmental sustainability of cyber physical production systems. Procedia CIRP 69:644–649

Thiede S (2021) Digital technologies, methods and tools towards sustainable manufacturing: does Industry 4.0 support to reach environmental targets? Procedia CIRP 98(2021):1–6

Thiede S, Turetskyy A, Loellhoeffel T, Kwade A, Kara S, Herrmann C (2020) Machine learning approach for systematic analysis of energy efficiency potentials in manufacturing processes: a case of battery production. CIRP Annals

U.S. Department of Energy, Office of Energy Efficiency and renewable energy, https://www.energy.gov/eere/amo/static-sankey-diagram-full-sector-manufacturing-2014-mecs

Chapter 9
Insights: Green Verbund

Uwe Liebelt

1. **BASF's highly successful Verbund concept is praised throughout the industry as it reduces raw material use, energy consumption and consequently costs and greenhouse-gas emissions. What are the key factors of your success?**

Among the over 250 manufacturing sites BASF is operating globally, six are called "Verbund" sites. These sites are characterised by two criteria: firstly, production structures that comprise entire chemical value chains, and secondly, the sites' multi-dimensional integration. Waste streams and by-products of one process are reused as raw materials for other processes; excess heat of exothermic reactions is recovered and fed into the central steam-grid to be used by other plants; raw materials and finished goods always find the optimal transportation mode as they can be seamlessly switched between pipeline, ship, train and truck. On top is a data integration layer allowing, for example, central steering of the onsite steam-and-power-grid. The efficiency driven by the Verbund can be measured: integrated manufacturing saves roughly €1 billion in cost and 6 million tons of CO_2 emissions per year.

2. **Ludwigshafen is considered a role model for digitalisation in the chemical industry. How did you manage the transformation?**

BASF's Ludwigshafen site is its largest Verbund site and at the same time, the largest integrated chemical manufacturing site owned by one company in the world. 200 plants connected by 2800 km of pipeline on a space of 10 km^2 producing roughly 8 million tons of products for our customers—these are some of the impressive numbers hallmarking the site.

As the largest Verbund site in our portfolio, Ludwigshafen was the logical choice for starting digitalisation—the most complex challenge combined with the biggest

U. Liebelt (✉)
BASF, Carl-Bosch-Str. 38, 67063 Ludwigshafen, Germany
e-mail: uwe.liebelt@basf.com

© Springer Nature Switzerland AG 2022
O. Inderwildi and M. Kraft (eds.), *Intelligent Decarbonisation*, Lecture Notes in Energy 86, https://doi.org/10.1007/978-3-030-86215-2_9

benefit potential. We commenced in 2014 when digitalisation in chemical operations was something entirely new. Of course, we had experimented with neural networks in the late 1990s like many others and in some high value product lines, model predictive control was in use. However, on a general scale, computing power and transmission speed per Euro had not been large enough before 2014 to handle the immense amount of data typical for chemical operations in affordable and user-friendly applications.

Today, digital tools have found their way into every corner of operations at the Ludwigshafen site. Around 300 Predictive Maintenance Machines are used as early warning systems for equipment performance losses or unplanned outages. A large variety of processes is driven towards highest efficiency by algorithms permanently optimising yield, conversion rate or product quality. The steam-and-power-grid is controlled and steered by three interacting AI systems. Overall margin optimisation across all product lines can be managed through a digital twin of the Verbund and the world's first autonomous heavy load vehicles for chemicals are driving 40-ton tank containers to and from the loading bays. Most importantly from my point of view, all these powerful tools are accepted and used by operators and engineers as we have spent as much energy on UX design and training as we have spent on developing algorithms.

3. **These sound like significant investments into Industry 4.0. Are they paying back?**

Whether we think about processes running outside their optimum, unplanned outages or inaccurate planning and forecasting—all those situations typically result in higher raw material intensity, higher energy consumption, higher CO_2 output and higher cost. AI in Industry 4.0 plays an important role in continuously driving plants towards higher efficiency and thereby reducing CO_2 emissions and cost. It is important to mention that the payback time of 4.0 investments usually is much shorter than a standard investment in operations. The reason is that the chemical industry is asset-heavy and capital intensive. So, if you optimise an expensive and complex fixed asset with comparably cheap data tools, the pay-out is almost immediate.

4. **Do you see even more potential for further efficiency improvement regarding CO_2 emissions through AI?**

Yes, I am deeply convinced we are just at the beginning of this. BASF supports the Paris Agreement on climate protection and the European target to become the first carbon–neutral continent globally by 2050. Notwithstanding an already significant usage of digital tools in operations, dissemination of AI within the Verbund will further increase. A different opportunity I see within the Verbund is shifting the overall optimisation target from cost towards CO_2 minimisation. Starting 2021, we will provide our customers with CO_2 footprints for every single BASF product they buy, so-called "Product Carbon Footprints" or PCFs. As PCFs will become an increasingly important competitive factor in the future, we will use AI to guide the Verbund towards higher CO_2 efficiency.

Steering towards greenhouse gas neutrality by 2050 will require more than mere additional efficiency gains. Instead, fundamental changes in the way the chemical industry produces energy and products will be required. A replacement of "grey" energy produced in fossil-fuel based power plants by "green" energy and the electrification of chemical processes is indispensable from my point of view. The resulting surge for green energy will not be compensable through efficiency advancements thus, demand for green power will massively increase. At the same time, the network of power producers, storage facilities, converters and users will become more diversified, decentralised and volatile. To control, steer and stabilise the much more complex power grid of the future, AI will be key—not only for the chemical industry.

Dr. Uwe Liebelt started his career with BASF in 1996 in the research department. Between 2005 and 2014, he led several Business Units and a Global Operating Division in the United States, Switzerland and Singapore. In 2015 he initiated BASF's digital journey, kick-starting project BASF 4.0. Since 2016, he has been the President of BASF's European Verbund sites, a role in which he, amongst other things, oversees digitalisation and transformation towards carbon–neutral operations.

Chapter 10
Green, Lean and Digital Transformation for Decarbonisation of Chemical Industries

Oliver Lade

Abstract This chapter assesses the status and potential of Chemical Industries' transformation towards more green, lean and digital operations and products. The reasons are discussed that apparently delay the more consistent application of Industry 4.0 methods. Industry best practices for identifying, implementing, and applying technologies and methods for the decarbonisation of chemicals over their entire lifecycle are delineated. Two main strategies are distinguished, each having their own approach and benefits potential: The game changers for blockbusters versus the incremental improvements for long-tail products. Typical valuable, albeit too often neglected use cases are shown and exemplarily briefly discussed. We believe that the 15 years ongoing stagnation of the specific energy consumption in Chemical Industries can be transferred into a decrease again as typical in the 1990s. By comparing with other similar industries we estimate the decarbonisation potential to be 15% in the next 10 years.

10.1 Introduction

Process industries (including chemical, pharmaceutical, industrial biotechnology, paper, cement, glass, food, and water) have made significant progress in sustainability and productivity in recent years. In the chemical industry, for example, resource and energy efficiency as well as process and occupational safety have improved steadily and remarkably through the development and application of better production processes. This was achieved primarily through better chemical and technical understanding and implementation at the production level. In both fields, classical technologies and methods of digitization such as (soft) sensors, modeling, and simulation, advanced process control, etc. play an important role.

O. Lade (✉)
Faculty of Computer Science and Business Informatics, Provadis School of International Management and Technology AG, Industriepark Höchst, 65926 Frankfurt am Main, Germany
e-mail: oliver.lade@doz-provadis-hochschule.de

© Springer Nature Switzerland AG 2022
O. Inderwildi and M. Kraft (eds.), *Intelligent Decarbonisation*, Lecture Notes in Energy 86, https://doi.org/10.1007/978-3-030-86215-2_10

The CO_2 footprint of chemicals[1] stems from various sources: From the energetic use of carbonized fuels in their manufacturing, but—more pronounced compared to other manufacturing industries—also from thermal recovery or from composting at the end of their lifecycle. Therefore, the use of renewable or recycled carbon sources has a positive impact with respect to the CO_2 footprint.

In industrialized countries, the Chemical Industries consume typically 10% of the overall energies (Calculation of the VCI (Verband der chemischen Industrie)) and 20% of the energies in the manufacturing sector (Calculation of the VCI (Verband der chemischen Industrie)). In Germany in 2015, the manufacturing of chemicals consumed twice as much energy (1,4 Terajoule) compared to metals (0,7 Terajoule), which was the second most energy-consuming industry sector (Bundesumweltamt: Branchenabhängiger Energieverbrauch des verarbeitenden Gewerbes 2018).

In the 1990s and early 2000s, the energy consumption of the Chemical Industries decreased at raising production rates, i.e. the specific energy consumption decreased by 50% in 14 years. However, from 2004 to 2017, the specific energy consumption remained at the same level (Calculation of the VCI (Verband der chemischen Industrie)) and in the same period, only a slight decrease of specific greenhouse gas emissions[2] of 10% (compared to the level of 1990) could be achieved.

Especially compared to other manufacturing industries, though, the specific energy efficiency in Chemical Industries lacks significant improvements since 2004 (Fig. 10.1).

Our working hypothesis of these observations is that traditional technologies have successfully been implemented, but improvements from advanced digitization and innovative technologies (Chan et al. 2019; Fleiter et al. 2019; Chiappinelli et al. 2020) have not been fully utilized. Chemical Industries are significantly behind other industries in these respects.

10.2 Green, Lean and Digital Transformation

10.2.1 Status and Current Obstacles

Green, lean and digital principles in Chemical Industries

The systematic implementation of lean principles (Fig. 10.2) in the production of process industries began around 1990—much later compared to industries with discrete manufacturing (Panwar et al. 2015; Floyd 2010). Lean manufacturing originally means incremental improvement in operating plants, but the principles can be adapted to the design of new products or plants (Staudter et al. 2013).

---

[1] In this article we refer to chemicals as pharmaceutical products, petro chemicals and intermediates, specialty chemicals and formulations, e.g. raw materials for plastics, paints, cosmetics etc. Accordingly, Chemical Industries also comprises pharmaceutical and life science companies.

[2] Energy related CO_2 emissions and N_2O emissions.

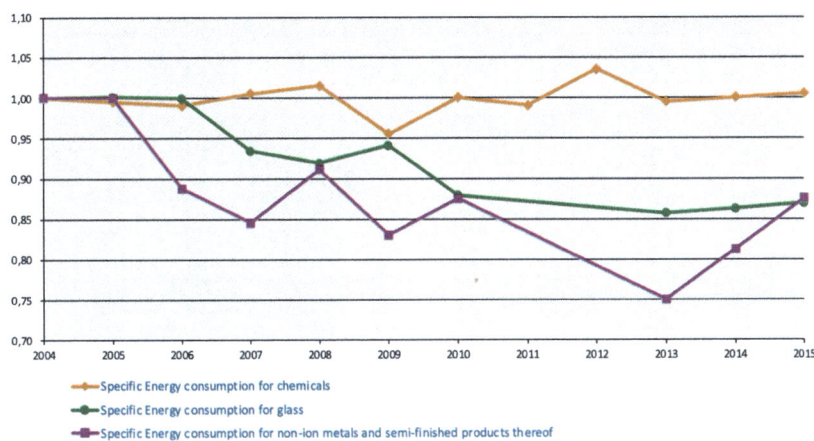

Fig. 10.1 Development of specific energy consumption for chemicals, glass and metals 2004 to 2015, Index 2004 = 1. Redrawn from Calculation of the VCI (Verband der chemischen Industrie), Bundesumweltamt: Branchenabhängiger Energieverbrauch des verarbeitenden Gewerbes (2018)

Fig. 10.2 Key governing principles of lean manufacturing (Capgemini Invent, Fraunhofer IGCV 2020)

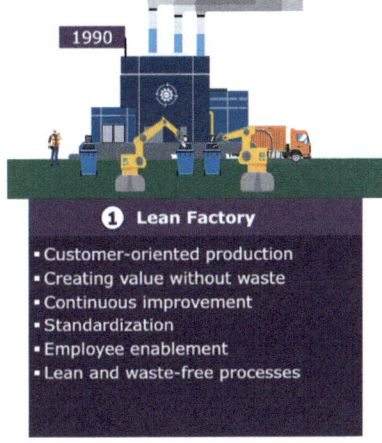

Many lean use cases benefit from the systematic application of data-driven digital methods to gain better transparency, controllability, and predictability of production processes (Fig. 10.3).

All activities within the factory of the future will be balanced with respect to economical efficiency, environmental impact and technological scientific progress (Capgemini Invent, Fraunhofer IGCV 2020) (Fig. 10.4).

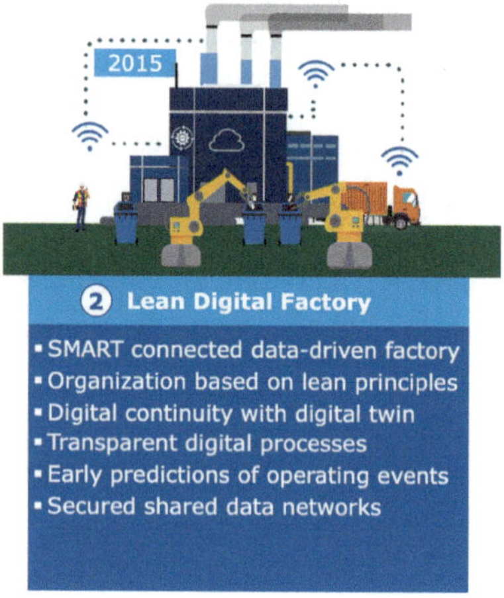

Fig. 10.3 Key governing principles of lean digital manufacturing (Capgemini Invent, Fraunhofer IGCV 2020)

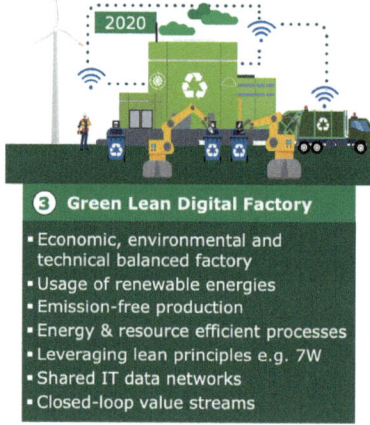

Fig. 10.4 Key governing principles of green lean digital manufacturing (Capgemini Invent, Fraunhofer IGCV 2020)

Current actual maturity in Chemical Industries

Due to the very long lifecycles of typical chemical plants, production processes very often are based on traditional know-how and experience. Typically, much data is recorded and stored, but the latter in scattered silos and without internal and mutual connection, thus difficult to access for a systematic evaluation.

When changing the recipe, the raw material vendor, or the product specification, e.g. due to regulatory changes, is required in such plants, then this is cumbersome to implement within a reasonable development effort and often does not lead to an optimal operation.

Also, an increase in plant capacity, e.g. through debottlenecking or a new investment, is often associated with considerable development effort, because a detailed process and plant understanding is missing or the bases for the scale-up are often no longer available.

Even where extensive know-how has been built up for recent plants in process development, including possibly process models, a consistent transfer of this knowledge over plant design, construction, and commissioning to production is not always given. Finally, even successfully transferring process knowledge from development to production level often lacks the validation of that knowledge with the help of widely available production data. However, this is not primarily a matter of technology, because in recent years big data analytics and numerous cloud-based offerings have developed an agile method toolset that enables application-specific implementation in very short sprint cycles in just a few weeks.

In the field of methodology, the classical-deterministic modeling and simulation of thermodynamics, kinetics, and processes (e.g., cycle times etc.) has been established for chemical process development. Even if the motto "No model, no plant" is not consistently maintained, modeling the critical process steps (reactors, thermal separation) is common practice. The same methodology is also used for overcoming capacity bottlenecks (debottlenecking).

However, only in parts of the current production an improvement in the key parameters is realized by systematic analysis of existing data and use of the insights gained. Even the lowest levels of the automation pyramid—the control of individual process parameters—show, according to a study (Van Doren 2008), a surprisingly little knowledge-based status: one-third of the control circuit is not in operation at all, another third is operated in the basic settings of the control parameters and only the last third has adapted control parameters.[3] Also in Bauer et al. (2016) wrong parameter setting is named as the most common problem in closed-loop performance.

Humans as a success factor

As with many change processes, it is largely the people who decide—willfully or subconsciously—on success or failure. It must be recognizable to them how digital transformation addresses their concrete questions on the shop floor. There is no one-fits-all strategy, every question requires individual approaches and solutions.

The White Paper "Professions 4.0—How Chemists and Engineers Work in Digital Chemistry" by the Association for Chemistry and Business, VCW (GDCh Section) identifies two major challenges.

"Many chemists and engineers want to be involved wherever possible and keep control in detail. Future competency-based job profiles and work content—along with 'sharing' tasks and 'handing over' individual responsibilities to other key competencies—are what make up most chemists and engineers. Emerging professions such as rather IT-oriented pure data scientists, more balanced chemistry computer scientists, and primarily chemical and business-oriented value chain managers as well as stakeholder managers in the environment of large projects are currently known as a concept, however, the belief in them as new colleagues with specialist skills is almost completely missing."

---

[3] Although 10 years old, the cited study still reflects at least qualitatively the current situation (as per our own project experiences).

A combination of top-down and bottom-up strategies promises most success when it comes to utilizing executive as well as shop-floor intelligence in order to achieve heuristic results and the highest acceptance throughout the whole organization (Fig. 10.5).

General success factors in digitalization projects

Capgemini's study on IT Trends (cross-German-industries, 2020) gives valuable hints for general success factors within digitalization projects (Fig. 10.6).

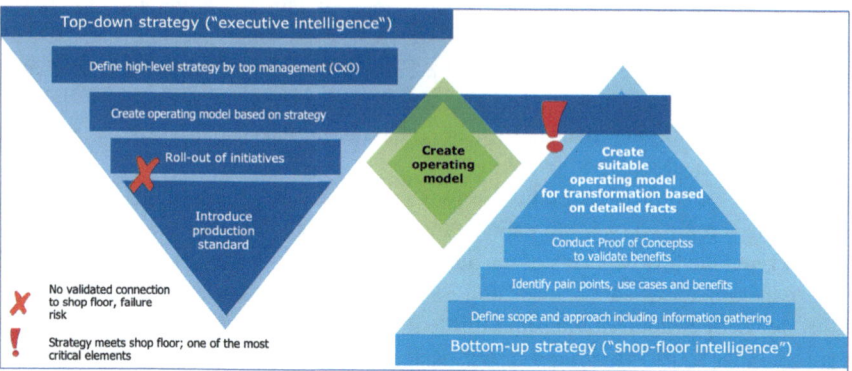

Fig. 10.5 Combining top-down and bottom-up for best results and acceptance

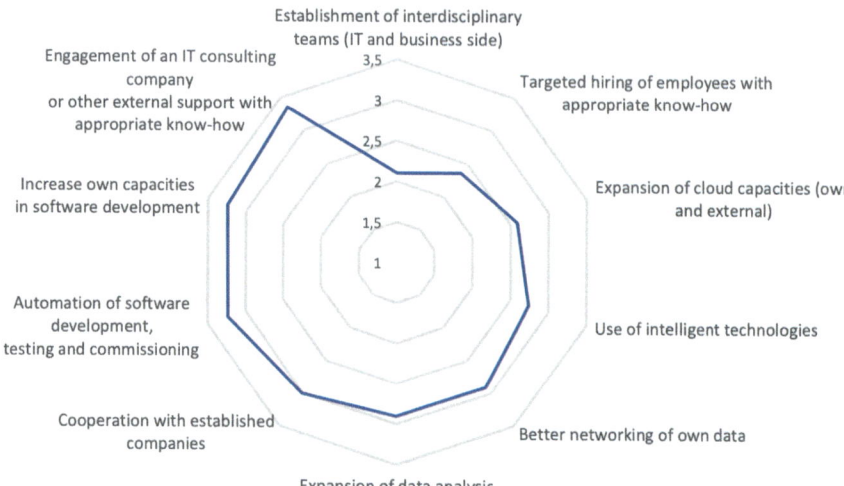

Fig. 10.6 How important were the following measures for the success of digitization in your company? Top 10 answers of all interviewees with digitization success (n = 96), mean values on a scale of 1 (very important), 5 (completely unimportant)

Game changers for blockbusters versus incremental improvement for longtail products

The chemicals production landscape is much more fragmented compared to e.g. automotives.[4] When it comes to product volumes and specific chemical engineering processes, very few base chemicals dominate and act as the basis for a wide variety of products and processes (Technology Roadmap 2013). Hence, in this article we will focus on a few thoughts towards game changers for blockbusters (2.3) and towards incremental improvement methods for the longtail products (2.4), respectively.

For blockbusters—as they have gone through long years of optimization—bold measures are needed to result in small relative changes, which due to the huge production amounts add up to significant absolute improvements.

Production processes for products of the second and third row provide much more relative improvement potential, however, a lot of such improvements must add up to become significant in total. Therefore, a systematic methodology combined with a proven best practice tool stack must be applied routinely and systematically in broad initiatives.

10.2.2 Chemistry 4.0 Tools and Methods

Lean principles

Lean manufacturing is foremost based on a culture where management takes care that the shopfloor working force is enabled to do what they strive for: Getting things done most efficiently, i.e. without wasteful and meaningless activities at a high level of quality and safety with the least effort and rework possible.

The ESOAR method (Fig. 10.7) guides the way towards the maximum lean shopfloor.

Digital Transformation

Methods and tools in digital transformation, e.g. Cyber-Physical-Systems (CPS) and Artificial Intelligence, have been summarized for the energy sector (Inderwildi et al. 2020). The same principle concepts apply in chemical industries. Digitalization should cover the whole product and process life cycle with integrated and seamless information availability (Fig. 10.8).

The currently highest unused potential of CPS in chemicals industries we see in big data analysis, the overarching goal of which is to support decision-making processes ultimately improving the profitability of a company and the sustainability of its products (Wachsen et al. 2015; Appel et al. 2016).

---

[4] E.g., BASF operates more than 360 production sites worldwide, among which are 6 so-called Verbund sites (such as Ludwigshafen, Antwerp, etc.), and achieves a revenue of 59 b€ in 2019. Daimler, on the other hand, has 40 production sites worldwide with revenues of 173 b€ in 2019.

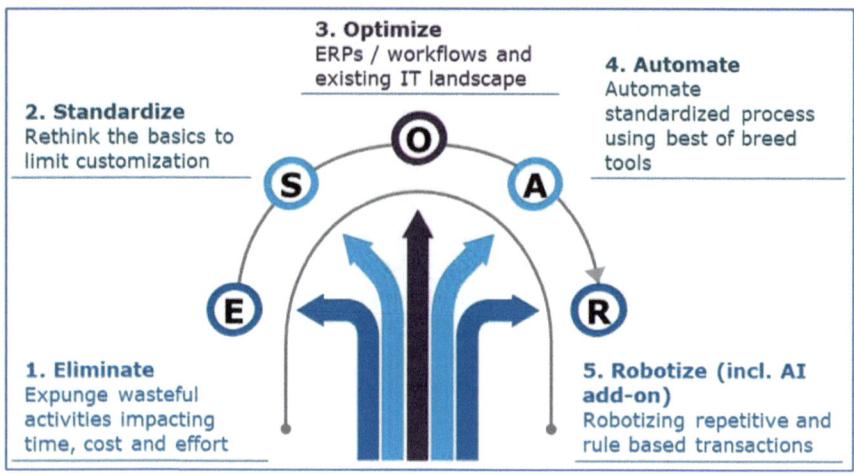

Fig. 10.7 The ESOAR method provides a staged approach to lean manufacturing (Capgemini Invent, Fraunhofer IGCV 2020)

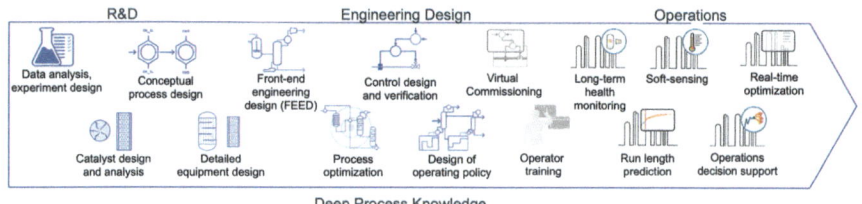

Fig. 10.8 Digitalization over the whole product and process life cycle (PSE Ltd.)

However, required existing data is often not available in one place and is not available in a common format; due to the mixing of batches and dwell times in the plants there is additional ambiguity. The history of a product must be traced back using model assumptions. Despite these challenges, it is worth structuring the existing information and bringing it into an evaluable form. The creation of a permeable infrastructure for the data is the most important success factor in data analysis for product and process optimization. A variety of commercial software solutions is available.[5]

The evolution stages of digital manufacturing are shown in Fig. 10.9.

---

[5] (a) www.osisoft.com;

(b) www.aspentech.com/products/aspen-infoplus21/;

(c) https://www.siemens.com/global/en.html;

(d) https://community.sap.com/topics;

(e) www.lims.de.

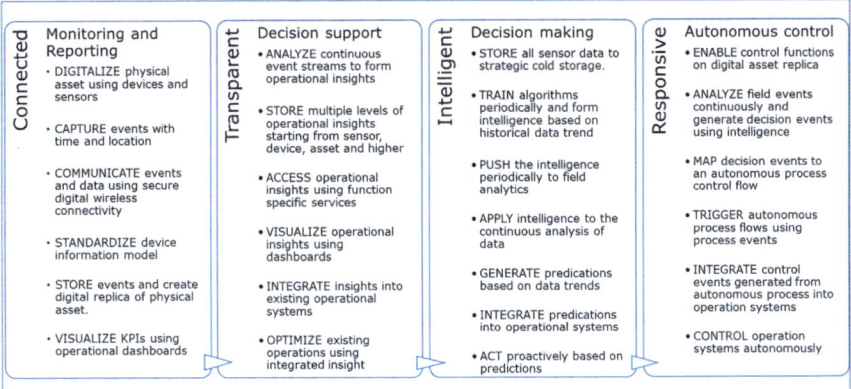

Fig. 10.9 Four maturity stages in digitized manufacturing (Capgemini Invent, Fraunhofer IGCV 2020)

Green Manufacturing

Based on a life-cycle assessment (Fig. 10.10) a prioritized list of pain points is generated and these can be addressed by ideation workshops and by defining corrective and mitigating actions or projects.

Fig. 10.10 Basic milestones of a life-cycle-assessment

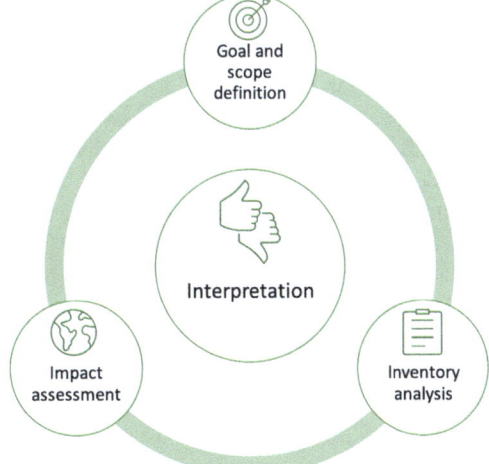

10.2.3 Opportunities for Carbon Emissions and Cost Reduction—Game Changers

The four most produced base chemicals are olefins (ethylene, propylene), ammonia, aromatics (BTX: benzene, toluene, xylenes), and methanol. Their production accounts for almost half of the total energy demand of chemicals industries, whereas the 18 most-produced chemicals in total account for almost two-thirds of the total production energy demand (Technology Roadmap 2013). One of the most CO_2 intensive chemical productions shall serve as an example of how process engineering innovations foster green game changing:

Other than e.g. in gas-to-liquid technologies such as the Fischer–Tropsch-process, in the industrial ammonia production the carbon from the methanol steam cracking is not bound in the final value-product, but released as CO_2. In large-scale ammonia production, however, the heat of reaction is used to produce steam, which - when consumed in other close-by sites in chemical parks - has a certain beneficial impact on the Global Warming Potential (GWP).

Two game changers are currently discussed—three if we consider a combination of both—, but they require further technology and logistic innovations: (1) The acquisition of hydrogen from electrolysis of water with hydropower would reduce the CO_2 footprint significantly, but would also yield much less steam. Therefore, this concept would suffice for smaller decentral units, because no steam sink is needed. (2) The replacement of the conventional catalytic Haber–Bosch reaction by a non-thermal ambient pressure plasma-assisted reaction fed by renewable energies.

In an extended end-to-end study (Anastasopoulou et al. 2020) the conventional process is compared with several scenarios of the above innovations in a distributed, local production business model. The process comprising renewable hydrogen, but conventional catalysts is comparable with respect to the GWP to the conventional Haber–Bosch ammonia process using hydrogen from fossil sources—but only when considering the steam credits in the fossil case. However, the availability of large amounts of steam does often not foster an energy-sensitive behavior.

This new route to ammonia exemplifies the electrification of chemistry—one of the break-through game changers for decarbonisation in Chemical Industries (Hessel 2015; Schiffer 2017; Barton 2020)—in two trends: Hydrogen/acetylene based New Verbund[6] (Wehberg 2015) and plasma-based processes.

Likewise, for a wider usage of CO_2 as a raw material, e.g. for methanol (Bukhtiyarova et al. 2017), an industry scale green hydrogen New Verbund or the direct activation of CO_2 by e.g. plasma-based processes (Ma et al. 2019) is inevitable.

---

[6] The Verbund concept is known as a major asset of BASF, which they define as follows (BASF Internet): "The driving principle of the Verbund concept is to add value through the efficient use of resources. At our Verbund sites, production plants, energy and material flows, logistics, and site infrastructure are all integrated."

Under New Verbund we understand the same principle, but with new, even more sustainable value chains.

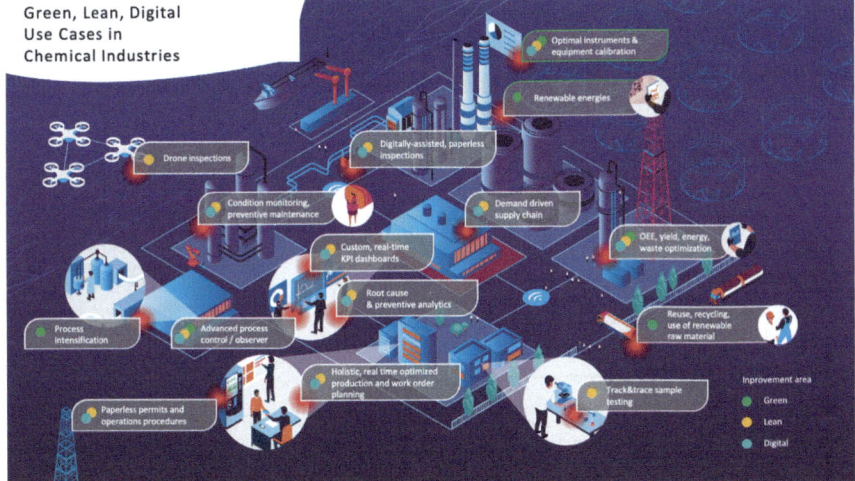

Fig. 10.11 Typical state-of-the-art green, lean and digital use cases of a smart chemical site

10.2.4   Opportunities for Carbon Emissions and Cost Reduction—Incremental Improvements

Compared to a best-in-class smart chemical site—typical state-of-the-art use cases are shown in Fig. 10.11—our experience from on-site end-to-end assessments is, that chemical production on the actual shopfloor level is far behind its potential. This is interestingly the case over all scales of productions and sites—and some few lighthouse examples should not be mistaken as the broad development status.

Lean

Current tangible and valuable, albeit not fully rolled-out technologies to cope with one or more of the seven types of waste defined in Lean are e.g. (1) demand-driven material requirements planning (DDMRP), (2) sales and demand planning with real-time optimization and (3) supply network optimization.

Digital

Typical ready-to-use and usually very beneficial levers in chemical production are (1) digital operator concepts for paperless, better informed and documented operations, inspections, and maintenance and (2) plant control tower concepts for tailored transparency about the status and suggested near-future optimized operations of the plant. Autonomous operations for further slight improvements are in principle possible, but currently still face high regulatory and cultural hurdles.

Green

Typical green levers in Chemical Industries are summarized in Fig. 10.12.

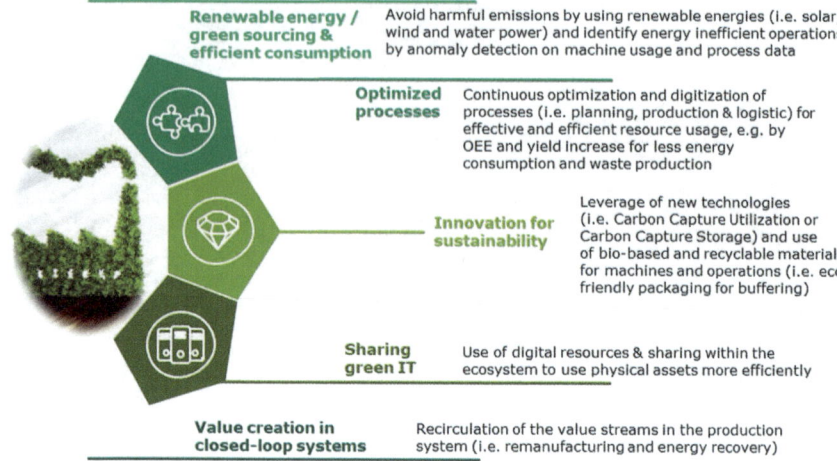

Fig. 10.12 Typical state-of-the-art green levers in Chemical Industries (Capgemini Invent, Fraunhofer IGCV 2020)

10.3 Conclusions

For a successful transformation of Chemical Industries to become more green, lean, and digital in order to better balance costs and greenhouse gas footprints we see these key strategic recommendations—adapted from Capgemini Invent, Fraunhofer IGCV (2020), Wehberg (2015):

1. Act fast—especially for new investments

   Given the long life cycles of chemical products and productions there is a long-lasting do-nothing impact, i.e. develop and apply innovations fast so that for the very next generation of new investments the optimal decarbonisation effects can be realized.

2. Leverage digitalization fully

   Many companies are focusing on reproducing existing processes as-is in an electronic fashion. Opportunities for developing new operating, supply-chain, and business models, respectively, are not being addressed comprehensively.

3. Don't go the Big Bang approach—unless for Game Changers

   The impact of Industry 4.0 within Chemical Industry will, for the most part, materialize in small steps and in the midterm, i.e. in an evolutionary fashion rather than a "Big Bang". However, for high-volume processes conventionally optimized for decades, an innovative game changer approach often is required, e.g. electrification of chemistry.

4. Utilize Big Data analytics to master complexity

   Data-based operating models leverage predictive analytics, resilient network structures as well as self-organization. Such operations are best prepared to anticipate new market developments.

5. Use Industry 4.0 as a facilitator

    Industry 4.0 also provides solutions for responding to other key trends as a facilitator. Examples are smart solutions for energy efficiency and individual offerings for changing customer behavior.

6. Explore potentials beyond technology

    Industry 4.0 needs to be viewed through business-strategic glasses in terms of efficiency and growth, i.e. tapping lean potentials and building green business cases.

7. Deploy a New Green Verbund

    Green portfolio development strategies question previous integrated supply chain models. Therefore, Verbund has to develop towards a New Green Verbund.

8. Include all relevant stakeholders to cope with complexity

    In order to secure competitiveness and sustainability in the future, key external and internal factors have to be considered in joint projects with all relevant stakeholders.

9. Chempark operators: Provide Industry 4.0 infrastructure

    Site operators are requested to provide proper infrastructure solutions for smart operations as well as on-site connectivity, e.g. by 5G industrial networks.

10.4   Outlook

Acting fast to leverage state-of-the-art green, lean, and digital methods and—if beneficial—innovative technologies together with the required policy (Chiappinelli et al. 2020), cultural and educational changes should enable Chemical Industries to follow the path of increasing specific energy efficiency—like other similar process industries—and to realize other decarbonisation potentials of in sum up to 15% over the next 10 years.[7]

Acknowledgements The Green-Lean-Digital Method and many of the diagrams in this chapter originate from a collaboration between Fraunhofer IGCV and Capgemini Invent. Involved in this development were—amongst others—Ralph Schneider-Maul, Niko Seeger, Johannes Metten, Tobias Wissing, Andrea Hohmann, Martin Rösch, Jana Köberlein Götz Wehberg was a valuable discussion partner, e.g. in the broad topics of Chemicals 4.0—cf. (Wehberg 2015).

References

Anastasopoulou A, Keijzer R, Patil B, Lang J, van Rooij G, Hessel V (2020) J Ind Ecol, 1–15
Appel J, Colombo C, Dätwyler U, Chen Y, Kerimoglu N (2016) CHIMIA Int J Chem 70(9):621–627
Barton JL (2020) Electrification of the chemical industry. Science 368(6496):1181–1211
Bauer M, Horch A, Xie L, Jelali M, Thornhill N (2016) The current state of control loop performance monitoring—a survey of application in industry. J Process Control 38:1–10

---

[7] For a much deeper analysis cf. (Fleiter et al. 2019).

Bukhtiyarova M, Lunkenbein T, Kähler K, Schlögl R (2017) Methanol synthesis from industrial CO_2 sources: a contribution to chemical energy conversion. Catal Lett 147(2)

Bundesumweltamt: Branchenabhängiger Energieverbrauch des verarbeitenden Gewerbes—www. umweltbundesamt.de/daten/umwelt-wirtschaft/industrie/branchenabhaengiger-energieverbr auch-des#der-energiebedarf-deutschlands (2018)

Calculation of the VCI (Verband der chemischen Industrie)—cf. www.vci.de/ergaenzende-downlo ads/energy-statistics-for-the-german-chemical-pharmaceutical-industry.pdf

Capgemini Invent, Fraunhofer IGCV, Green Lean Digital Factory—Sales Deck, July (2020)

Chan Y, Petithuguenin L, Fleiter T, Herbst A, Arens M, Stevenson P (2019) Industrial innovation: pathways to deep decarbonisation of industry. Part 1: Technology analysis, ICF and Fraunhofer (2019). https://ec.europa.eu/clima/sites/clima/files/strategies/2050/docs/industrial_innovat ion_part_1_en.pdf

Chiappinelli O, Erdmann K, Gerres T, Haussner M, Juergens I, Neuhoff K, Pirlot A, Richstein JC, Chan Y (2020) Industrial Innovation: Pathways to deep decarbonisation of Industry. Part 3: Policy implications, ICF and Fraunhofer (2020), https://ec.europa.eu/clima/sites/clima/files/strategies/ 2050/docs/industrial_innovation_part_3_en.pdf

Fleiter T, Herbst A, Rehfeldt M, Arens M (2019) Industrial innovation: pathways to deep decarbonisation of industry. Part 2: Scenario analysis and pathways to deep decarbonisation, ICF and Fraunhofer (2019), https://ec.europa.eu/clima/sites/clima/files/strategies/2050/docs/industrial_i nnovation_part_2_en.pdf

Floyd RC (2010) Liquid lean: developing lean culture in the process industries. Taylor and Francis, New York

Hessel V (2015) Electrification of chemistry—what is the synergy between plasma synthesis and chemical plant modularization? Green Process Synth 4:257

Inderwildi OR, Zhang C, Wang X, Kraft M (2020) The impact of intelligent cyber-physical systems on the decarbonization of energy. Energy Environ Sci 113:32–52

Ma X, Li S, Ronda-Lloret M, Chaudhary R, Lin L, van Rooij G, Gallucci F, Rothenberg G, Shiju NR,·Hessel V (2019) Plasma assisted catalytic conversion of CO_2 and H_2O over Ni/Al_2O_3 in a DBD reactor. Plasma Chem Plasma Process 39:109–124

Panwar A, Nepal BP, Jain R, Singh Rathore AP (2015) On the adoption of lean manufacturing principles in process industries. Prod Plan & Control: Manag Oper 26(7):564–587

PSE Ltd.—A Siemens Business

Schiffer ZJ, Manthiram K (2017) Electrification and decarbonization of the chemical industry. Joule 1(1):10–14

Staudter C, Hugo C, Bosselmann P, Mollenhauer J-P, Meran R, Roenpage O, Lunau S (eds) (2013) Design for six sigma + lean toolset, management for professionals, 2nd edn. Springer, Berlin

Technology Roadmap—Energy and GHG Reductions in the Chemical Industry via Catalytic Processes (2013): https://dechema.de/dechema_media/Downloads/Positionspapiere/Industria lCatalysis/Chemical_Roadmap_2013_Final_WEBcalled_by-dechema-original_page-124930-original_site-dechema_eV-view_image-1.pdf

Van Doren V (2008) Advances in control loop optimization. Software takes users from simple tuning to plant-wide optimization. Control Eng

Wachsen O, Bacher V, Geisbauer A, Appel J, Niebuhr V, Lade O (2015) Chem Ing Tech 87(6):1–12

Wehberg GG (2015) Chemicals 4.0—industry digitization from a business-strategic angle. www. Chemicals40.com

Chapter 11
Insights: Process Automation

Eckard Eberle

1. **As the CEO of Process Automation at Siemens, what role does digitalisation play in the decarbonisation of the process industries?**

Our digital solutions follow a clear purpose, which is to address the challenges our customers are facing in meeting their sustainability goals—for example, with regard to energy and material consumption, waste and water usage. At the same time, digitalisation helps our customers to shorten their time to market and to use their production facilities more flexibly. A true win–win situation. Our foremost enabler of digitalisation is our digital twin platform, which virtually represents a product, production, process or performance. In the design phase we can test solutions, make adjustments virtually and ensure compliance with environmental protection directives. During operations and with services, the digital twin assists in identifying potential for improvement.

2. **To what extent does Siemens make use of AI/machine learning today and what is the role of these technologies in decarbonisation?**

Siemens already makes broad use of AI and machine learning to create effective solutions for energy-intensive process industries, such as chemical, glass and water. Data-driven optimisation of membrane and distillation-based separation processes led to a 17% reduction of energy consumption in a chemical plant. In the water sector, 5–8% energy savings are possible with the optimisation of pump schedules and intelligent automation (Chap. 13). The more data that's available, the more improvements can be made using AI-based tools. In addition, we see the trend to hybrid models combining AI with model-based solutions (Chap. 10). One very good example comes from our acquisition of PSE with gPROMS process modelling tools.

E. Eberle (✉)
Process Automation Business Unit, Siemens AG, Munich, Germany
e-mail: eckard.eberle@siemens.com

© Springer Nature Switzerland AG 2022
O. Inderwildi and M. Kraft (eds.), *Intelligent Decarbonisation*, Lecture Notes in Energy 86, https://doi.org/10.1007/978-3-030-86215-2_11

With this approach, the results are even more precise and help our customers to improve the carbon footprint of their own operations.

3. How will the process industries implement the digital transformation and how will Siemens support this process?

Our extensive installed base and domain expertise in the process industries are our primary assets for supporting our customers. Because many topics relating to the process industries are very similar, we are able to transfer know-how gained in one sector to another, taking into account that each customer is different and that their needs are highly individual. Therefore, we recommend starting digitalisation projects with a consulting phase together with the customer. That serves as a basis to identify the requirements and goals for optimisation and to introduce an ideal end-to-end solution. At the moment a big focus is on modular production, which allows our partners to adapt to fast-changing trends and needs such as those now being experienced in pharma. The key drivers here are flexibility and correctly-sized production facilities.

4. This decade will see many transitions in almost every area of life. How do you and Siemens as a company support the ambitious goals in decarbonisation?

Siemens' overall aim is to be CO_2 neutral in all our factories and buildings, for our own operations, by 2030. To achieve this goal, Siemens Digital Industries is implementing and investing in according CO_2 reduction measures as a part of the Siemens Carbon Neutral Program. This is just one aspect of our commitment to a greener future. When it comes to our products, we estimate that less than 10% of the potential to reduce greenhouse gas emissions can be achieved during production, and the remaining 90% during usage. Technologies like AI and machine learning help adjust automation to keep emissions from use as low as possible. Customers can reduce operating costs in the short term through lower electricity consumption, and better fulfill their own ambitious climate targets in the long term.

Eckard Eberle is the CEO of Process Automation at Siemens AG since October 2014. Previous positions include CEO of Industrial Automation Systems, Head of the Control Components Business Segment, Head of the Low Voltage Transformer Business, and Siemens Management Consulting. Eckard's key priority is to leverage the digital transformation to create real value for his customers and ensure their competitiveness. He brings his dedication to technology and passion for innovation to shape the future of process automation. Eckard firmly believes that technology can make the world a better place to live and work.

Chapter 12
Intelligent Electricity Grid Management

Mario Baumgartner and Andreas Ulbig

Abstract Electricity plays a significant role for providing energy to society and enabling economic wealth. Electricity production today has a tremendous climate impact, corresponding to 40% of global GHG emissions. But electricity as an energy carrier can also be the enabler for an effective and large-scale decarbonization of both the mobility sector, by means of electric cars, and the heat sector, by means of heat-pumps. The first challenge in decarbonization is the integration of volatile RES production as world electricity generation increasingly shifts from fossil sources such as coal and gas towards renewable sources such as wind and solar power. The second challenge is the provision of the rising electricity demand due to electrification and the decarbonization of other sectors. The Smart Grid concept provides a framework to develop the electricity infrastructure's environmental friendliness, safety, reliability, and sustainability. At the center of the Smart Grid concept is an intelligent energy management of distributed electricity generation, delivery and consumption. Intelligent energy as well as grid management, however, requires transparency, especially on the side of the low voltage distribution grid. The rapidly rising capabilities as well as cost decreases of information and communication technologies (ICT) and artificial intelligence methods enable an improved grid transparency and numerous opportunities for grid analytics. In the following, this article discusses and showcases the potential of electricity grid analytics to provide new insights for decarbonization.

12.1 Introduction

Electricity as an energy carrier is already a significant source for providing energy to society and sustaining economic wealth. In addition, global electricity production is projected to increase to 44,000 terawatt-hours by 2050 (Statista 2020). This trend underlines the importance of the electricity sector for tackling climate change and decarbonization (Davis et al. 2018).

M. Baumgartner (✉) · A. Ulbig
Adaptricity AG, Hohlstrasse 190, CH-8004 Zurich, Switzerland
e-mail: mbaumgartner@adaptricity.com

© Springer Nature Switzerland AG 2022
O. Inderwildi and M. Kraft (eds.), *Intelligent Decarbonisation*, Lecture Notes in Energy 86, https://doi.org/10.1007/978-3-030-86215-2_12

In recent years, the emphasis of decarbonization was mostly on electricity production by replacing carbon-intensive power stations through renewables. Furthermore, rising electrification and hence the decarbonization of other sections, i.e. heat and transport, intensifies electricity demand and increases the need for a more flexible operation of electricity infrastructure.

Both challenges, volatile RES production and flexible infrastructure, require handling losses in transmission and distribution. Established over decades, the grid has certain inefficiency and hence surges losses through older inadequate assets, highly loaded network, or even the amount of power flowing changes. Such losses can be up to 30% depending on the country (OECD, IEA 2018).

To achieve a lasting reduction of losses and consequently CO_2 emission, electricity grids currently experience a transformation in technology, processing of data, and mindset. The Smart Grid vision emphasizes being more economical and environmentally sustainable. By enforcing actively efficient and effective management, objectives are the operation of new and improved equipment, integration of volatile decentralized generation and storage units, reduction of transformational steps, and power flow optimization (Failed 2017; Pournaras et al. 2017; Amleh et al. 2018). Such management requires transparency on all voltage levels, which is achieved by the handling of various data. Thus, Smart Grids utilize synergies between electrical engineering and advanced methods in information technology and data analytics. Eventually, digital solutions improve efficiency and reduce GHG emissions based on extended insights for decision-making.

12.2   Electricity and GHG Reduction

Electricity production has a significant impact on GHG emission reduction. Between the years 2000 and 2017, electricity production increased by about 70% reaching a total of close to 26,000 terawatt-hours (see Fig. 12.1) (IEA, World 2019). The world's

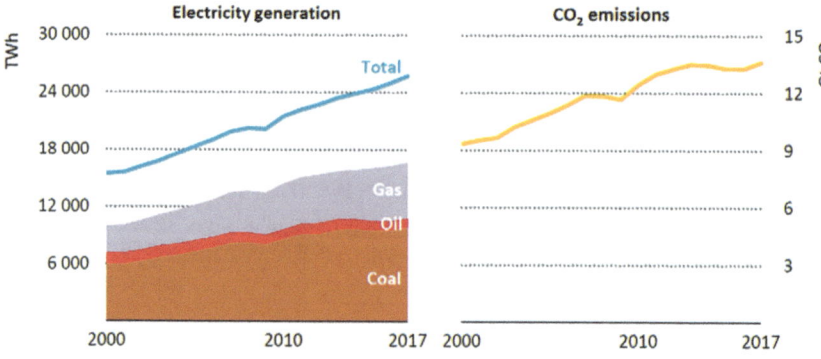

Fig. 12.1 World electricity generation and related CO_2 emissions 2000–2017 (IEA WEO 2018)

electricity generation mainly originates from GHG intensive sources, i.e. combustible fuels such as coal and gas. In 2017, the share of carbon-intensive sources was around two thirds.

Renewables, in particular solar and wind power gained traction and grew significantly since 2010 and global investment into renewable power generation has for several years been significantly outstripping investments in fossil-fueled generation (IEA, World 2019).

In terms of output, renewables still play a minor role in the electricity sector, but they are the lever to reduce GHG through a faster integration of renewables into the electricity grid. A higher velocity is critical for two reasons: (a) to cover future demand and hence to positively influence the carbon footprint, and (b) to replace existing fossil-fueled electricity production.

Digitalization of the electricity grid, including novel technologies in hardware, software, bidirectional communication, and methods in artificial intelligence, will be the key enabler for effective RES integration, and gives rise to advanced data analytics.

12.2.1 Smart Grids and Factors for Energy Savings

A Smart Grid is an evolutionary step for the electricity network and integrates subsystems and services governed through an intelligent management. The inherent grid management layer increases the infrastructure's environmental friendliness, safety, reliability, and sustainability (Santacana et al. 2010). Smart Grids also include self-regulating and self-healing attributes (Pournaras et al. 2017; Amleh et al. 2018) allowing real-time response to service interruptions or fault incidences.

The challenge for stable grid operation, and with it the potential for power interruptions, is the integration of volatile RES production, becoming less predictable, and simultaneously sustaining the stability of the electricity grid which relies on high predictability. Management must avoid supply and demand peaks. Both situations cause inefficiency and lead to electricity waste or usage carbon-intensive supply.

Balancing the grid, including batteries, is necessary for robustness, but must be orchestrated adequately. Another important factor for improving direct saving is line losses which derive from the conductor during the transmission and distribution phase. Smart Grids reduce transformation steps and save electricity through a decentralized production and reduced distribution distance, even so far as to have an on-site closed cycle. An effect of decentralization is the possibility to split the grid into virtual cells managing distribution efficiency.

In a Smart Grid, the electricity infrastructure contributes to GHG emission reduction through a holistic life cycle waste management (Wang et al. 2014; Ling and Song Anh Nguyen 2013). Aspects like lifetime expectancy and current and future fit of replacement of infrastructure require an in-depth insight for sophisticated decision-making. Especially, a lack of information in the design phase can increase the amount of waste (Wang et al. 2014). The Smart Grid relies on an advanced asset management enriched with utilization information, e.g. lifetime expectancy based on voltage data. Such data combinations allow stakeholders to learn and adapt behavior towards waste management (Osmani et al. 2008), which is an important factor for waste minimization (Poon et al. 2004).

12.2.2 Transparency as Key for Reducing Carbon Emission

Direct and indirect savings of electricity are essential for loss minimization and hence reducing GHG emissions. Regardless of the upcoming activities, the first step towards a sophisticated Smart Grid is an improved transparency in low voltage network for which the foundation is grid measurement data and, based on this, grid analytics. As Smart Grid represents the future architecture of the electricity network, transparency in the current situation is far from ideal. Grid transparency and with it the possibility for even simple analytics reduces quickly to zero when going from higher to lower voltage levels (see Fig. 12.2).

The development of the electricity infrastructure and its operation processes was made over decades by incrementally adding grid elements one at a time. Grid transparency only played a minor role and grid measurements were only recorded when needed. For example, the high voltage lines are tightly monitored because they represent the backbone of the electricity grid within and between countries. Transparency on the middle and low voltage grid levels, however, is neglected reprehensibly. Especially, the low voltage network is still operated as a black box and grid measurements are most often not available for either grid planning or grid monitoring. The utmost efforts are made when grid incidents occur, providing a sporadic glimpse into the actual grid status by dispersed and often only temporary metering.

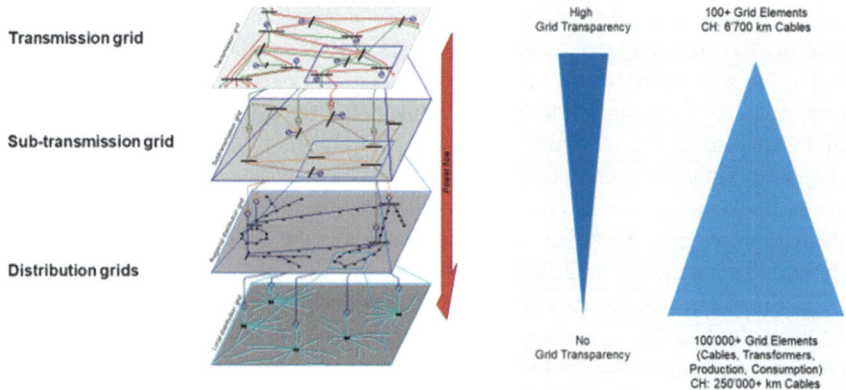

Fig. 12.2 Grid transparency significantly reduces from higher to lower voltage levels

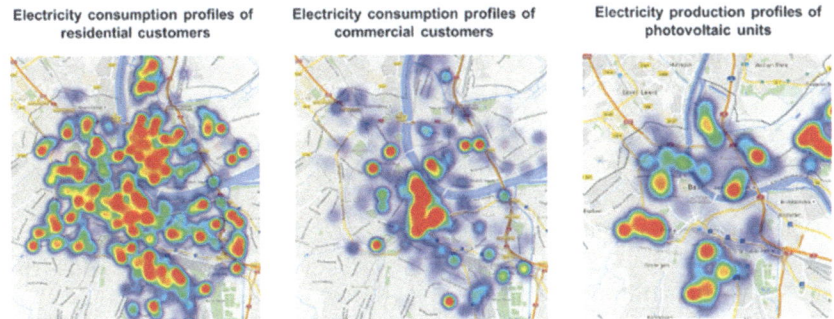

Fig. 12.3 Real-world examples of smart meter data analytics, city of Basel (Status in 2016)

At the low voltage level, more frequently and holistic monitoring through Smart Meter rollouts are introduced to improve data availability. The result is an evasive amount of measurement data requiring sophisticated handling as well as advanced analytics methods, as illustrated for the City of Basel (see Fig. 12.3). Such data amounts easily exceed human capabilities for digesting information adequately and hence human engineering skills must be supported by automated analysis insights. The advances in grid simulation and analytics (i.e. AI), facilitate modern planning and operation and reduce continuously ICT costs and satisfy regulatory requirements to justify investments and lifetime costs (MacCallum and Moore 2019). Eventually, digitalization of the electricity grid triggers potential for GHG emission reduction, which are shown by two real-life showcases.

12.3   Examples for Advanced Grid Analytics

Dealing with massive amounts of measurement data is as such already a complex task for humans. This task is further complicated in the case of a spatially highly distributed infrastructure like the electricity grid. Hence the need for automated data pre-procession and pre-analysis as well as intuitive visualizations, which enables a more effective and fruitful human–machine interaction, thus allowing humans to better grasp the inherent complexities of electricity grids and enabling effective decision making.

In the following two real-world showcases from Adaptricity's analytics projects with European distribution grid operator are presented.

12.3.1 Distribution Grid Monitoring for Loss Identification and Minimization

Adaptricity has helped the local grid operator of the Principality of Liechtenstein, Liechtensteiner Kraftwerke (LKW), to set up a continuous distribution grid monitoring system for its entire low- and medium-voltage grid (Fig. 12.4).

Continuous monitoring of the distribution grid infrastructure allows for better insights into the actual grid status and is a key enabler for more efficient grid operation and allows better utilization of the existing grid infrastructure.

The two main use-cases and benefits are the identification of existing grid bottlenecks, which enables line loss reduction, and an improved grid integration of RES generation.

The benefits of continuous monitoring for distribution grid operation are thus two-fold:

- Direct benefits in the form of more energy-efficient grid operation through a proactive reduction of line losses. In the case of LKW a plausible estimate is that up to 10% of the existing grid losses could be avoided in the future.
- Indirect benefits by enabling an increased hosting capacity of RES units and a fuel shift in the heating and mobility sector via heat pumps and electric cars. In the case of LKW, a plausible estimate is that up to 20% higher installed capacities of both RES generation and additional loads can be approved compared to traditional and necessarily more conservative grid connection procedures, when closely monitoring the physical limitations of the grid infrastructure.

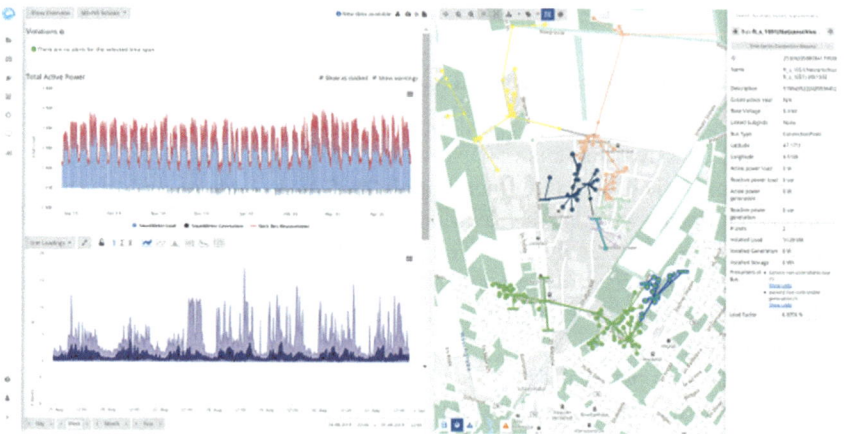

Fig. 12.4 Distribution grid monitoring on low- and medium-voltage level

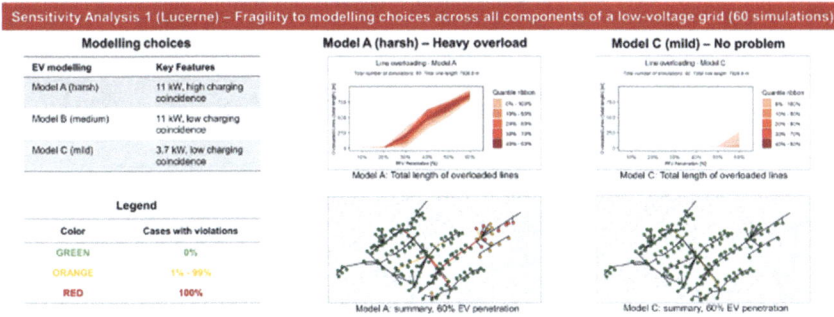

Fig. 12.5   Monte-Carlo-based grid impact analysis of increasing EV loading

12.3.2   *Predicting Grid Impacts from Decarbonization Trends in Cities*

In another recent grid analytics project, Adaptricity has helped the utility of the City of Lucerne (EWL) to explore grid development trends and identify potential future grid challenges in their jurisdiction.

An in-depth analysis of impacts from increasing load demand due to de-carbonization trends in both the mobility and heating sectors from new load types such as electric vehicles and heat pumps has been accomplished.

Future grid challenges caused by a rise of either electric cars or heat-pumps were robustly identified by using Monte-Carlo-based scenario exploration of dozens to hundreds of different plausible future grid setups. Analyzed grid challenges for these future grid setups were identified by looking at key grid operation metrics such as line overloading, voltage problems and transformer overloading.

Interestingly, for both types of new loads, a clear analysis result emerged that not all parts of the considered sections of the city's electric distribution grid are affected in the same way. It is clearly visible for both the case of rising electric vehicle charging, Fig. 12.5, and increasing heat-pump installations, Fig. 12.6, that certain parts of the electric network are more susceptible to higher peak loading than others.

This finding, in turn, gives rise to optimized grid investment strategies, allowing to pin-point typically limited funds to weaker network sections and/or otherwise neuralgic parts of the grid infrastructure.

12.4   Conclusions and Outlook

A more active grid management is needed for effectively managing volatile electricity production and demand peaks without overloading the grid infrastructure.

Fig. 12.6 Monte-Carlo-based analysis of grid impacts from increasing heat pump penetration in city neighborhood (5, 15, 25 and 35% penetration)

Building a Smart Grid, or at least a smarter grid is, due to the vast dimensions of the electricity grid infrastructure—the largest man-made invention—an evolutionary process, which starts naturally with improving grid transparency by advanced monitoring as well as intuitive grid analytics functions for this highly complex infrastructure.

The Smart Grid concept with its inherent grid operation and energy management function, allows a higher utilization level of the physical grid infrastructure without impairing grid stability. Advanced grid analytics functions such as grid monitoring and grid impact assessments will be more widely available in the coming years and provide a range of benefits.

In the case of wide-spread grid monitoring, particularly on the low-voltage level, there are direct benefits, such as more energy-efficient grid operation through an active reduction of line losses by 10%, and indirect benefits, such as enabling an increased hosting capacity of RES units, up to 20% more compared to conventional, more conservative grid planning.

In the case of detailed grid impact assessments, potential future grid bottlenecks can be identified more accurately and earlier. This allows grid operators to better prepare themselves for future challenges and direct infrastructure investments to where they are needed most and improve the cost–benefit ratio of grid investments.

In addition, a reliable electricity grid infrastructure that can both accommodate higher RES generation shares and additional load demand, permits a more rapid fuel shift from fossil combustible fuels towards carbon–neutral, renewable electricity in the heating and mobility sector via heat pumps and electric cars.

Acknowledgements The authors would like to acknowledge Adaptricity's project team, namely Nicolas Stocker, Damiano Toffanin, Janis Münchrath, Simon Schneller, Diren Toprak, Jochen Stiasny, Thierry Zufferrey and Antoine Gaillard, for their work on numerous grid analytics projects with grid operators over the last years including the show-cases presented here.

References

Amleh N, Al-Muhaini M, Djokic S (2018) Smart restoration for improved reliability of microgrids with renewable energy sources. In: 2018 IEEE PES innovative smart grid technologies conference Europe

Calatayud B, Candela A, Capelo J, Hilpert S, Liska J, Radovic O, Poljak M, Westermann M (2017) CEER report on power losses, Council of European Energy Regulators (CEER)

Davis JS et al (2018) Net-zero emissions energy systems. Science 360:6396

IEA, World Energy Outlook 2019, 2019. https://www.iea.org/reports/world-energy-outlook-2019

IEA/OECD (2018) World Energy Outlook 2018. https://www.iea.org/reports/global-energy-co2-status-report-2019/emissions. ISBN: 978-92-64-30677-6

Ling FYY, Song Anh Nguyen D (2013) Strategies for construction waste management in Ho chi minh city, vietnam. Built Environ Project Asset Manag 3(1):141–156

MacCallum JC, Moore G (2019) Drastic falls in cost are powering another computer revolution. Technol Q

OECD/IEA (2018) Electric power transmission and distribution losses (% of output). https://data.worldbank.org/indicator/EG.ELC.LOSS.ZS?end=2014&start=2014&view=map

Osmani M, Glass J, Price ADF (2008) Architects' perspectives on construction waste reduction by design. Waste Manag 28:1147–1158

Poon CS, Yu ATW, Wong SW, Cheung E (2004) Management of construction waste in public housing projects in Hong Kong. Construct Manag Econ 22(7):675–689

Pournaras E, Yao M, Helbing D (2017) Self-regulating supply–demand systems. Futur Gener Comput Syst 76:73–91

Santacana E, Rackliffe G, Tang L, Feng X (2010) Getting smart. IEEE Power & Energy Mag, 41–48

Statista (2020) Projected electricity generation worldwide from 2018 to 2050, by energy source (in trillion kilowatt hours), 27 June 2020. https://www.statista.com/statistics/238610/projected-world-electricity-generation-by-energy-source/

Wang J, Li Z, Tam VWY (2014) Critical factors in effective construction waste minimization at the design stage: a Shenzhen case study China. Resour Conserv Recycl 82:1–7

Chapter 13
Impact of Digital Transformation on Carbon Emissions Reductions in the Water Industry

Haifa Beji and Markus Lade

Abstract This chapter discusses how digital transformation can help reduce carbon emissions in the water industry. It highlights the two measures with the greatest decarbonization potential, namely energy efficiency and water efficiency, and describes how digital solutions can increase their impact at each stage of the water value chain. On the energy efficiency front, the chapter highlights how through process simulation and network optimization of desalination and wastewater treatment, as well as by adequate distribution network planning and operation, demand for energy and related carbon footprint can be reduced. On the water efficiency side, the chapter explains how digital solutions can decrease water losses in pipelines and therefore save on the water as a scarce resource and the associated energy-intensive production and distribution.

13.1 Introduction

There can be no doubt that water is the most precious resource on earth. Yet, in a world where new megatrends are emerging, the water sector is faced with the dual challenge of reliably providing potable water and wastewater services for a continuously growing population, as well as tackling the impact of climate change. Satisfying the rising demand for water is energy-intensive, whether it is for desalination, water treatment, distribution or wastewater handling. It has been estimated in the World Energy Outlook (WEO) that the water sector accounted for 4% of the global electricity consumption (IEA 2016). This makes the water sector also carbon-intensive, with the energy supply responsible for 3–5% of the global carbon dioxide emissions (WaCCliM 2014). The study foresees an increase in energy consumption in the next years, due to the increasing need for desalination plants and wastewater treatment plants. Additionally, there are significant embodied emissions, which the water companies are responsible for and which are relevant to the sector's greenhouse gas

H. Beji · M. Lade (✉)
Siemens AG-Digital Industries, Siemenspromenade 3, 91058 Erlangen, Germany
e-mail: markus.lade@siemens.com

© Springer Nature Switzerland AG 2022
O. Inderwildi and M. Kraft (eds.), *Intelligent Decarbonisation*, Lecture Notes
in Energy 86, https://doi.org/10.1007/978-3-030-86215-2_13

(GHG) balance. Wastewater treatment has for instance methane and nitrous oxide as by-products.

In their aim to meet the Sustainable Development Goals (SDGs) related to global GHG emissions reduction (SDG No. 13: Climate action) as well as related to sustainable management of water (SDG No. 6: Clean water and sanitation), cities have to rethink the way they are engineered considering the synergies between the different sectors (namely energy, transport, water and wastewater) and harnessing the benefits of the extraordinary technological advances. In fact, and taking the water and wastewater sector as an example, smart digital solutions are not only adopted to increase energy efficiency and reduce GHG emissions, but also to help save water resources. This chapter focuses on highlighting the role of technological innovation in reducing the water sector's carbon footprint. The chapter begins by examining the water cycle and identifying the main sources of carbon emissions. The next section looks into the potential of digitalization and artificial intelligence (AI) levers to achieve smart decarbonization in water and wastewater. Finally, the impact on emissions and costs in the future decades is investigated.

13.2   Carbon Emissions Reductions in the Water Industry

13.2.1   Water Supply and Wastewater Treatment Value Chain

The water cycle in Fig. 13.1 starts by sourcing through pumping from natural sources, i.e. surface water and groundwater, desalination or water treatment plants. Water is then conveyed from the sources to the end users. Finally, discharged water is collected and delivered via wastewater networks to the wastewater treatment plants. As depicted in Fig. 13.2, the overall energy consumed throughout the water cycle reaches 120 Mtoe (Million Tonne of oil equivalent) per year, which is almost equivalent to the entire yearly energy demand of Australia.

The carbon footprint of the water and wastewater sector is driven by various key factors, mainly

(1)   the energy-intensive desalination process
(2)   non-optimal pump operations in distribution networks
(3)   water losses (e.g. bursts, leakages) in distribution networks
(4)   the energy-intensive wastewater treatment plants.

As the ground- and freshwater resources are becoming scarcer and slower to replenish, seawater desalination is playing an increasingly important role. According to a 2019 World Bank study, desalination plants are suppling over 300 million people worldwide (The World Bank Group 2019). There are two main desalination technologies for large-scale water production: thermal processes and membrane methods. Thermal desalination consists in evaporating water and subsequently condensing it

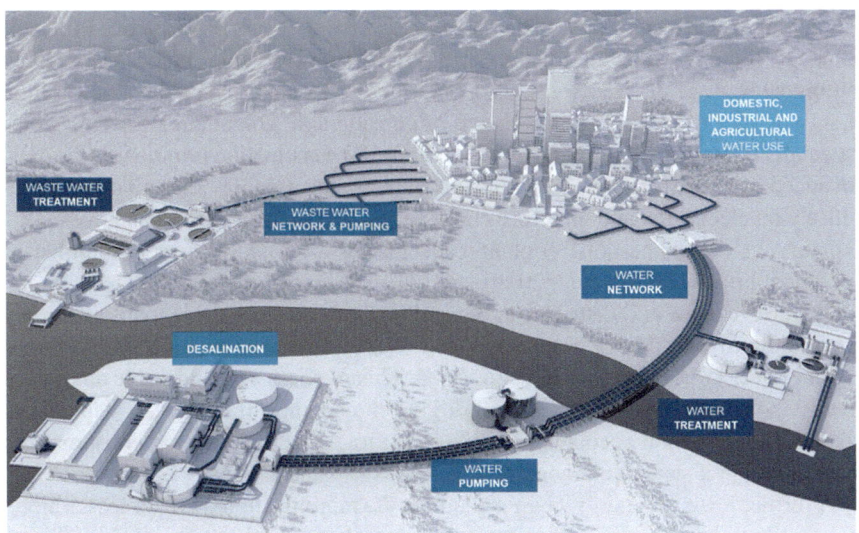

Fig. 13.1 Water cycle (Siemens AG, Germany)

The amount of energy consumed in the water sector is almost equivalent to the entire energy demand of Australia!

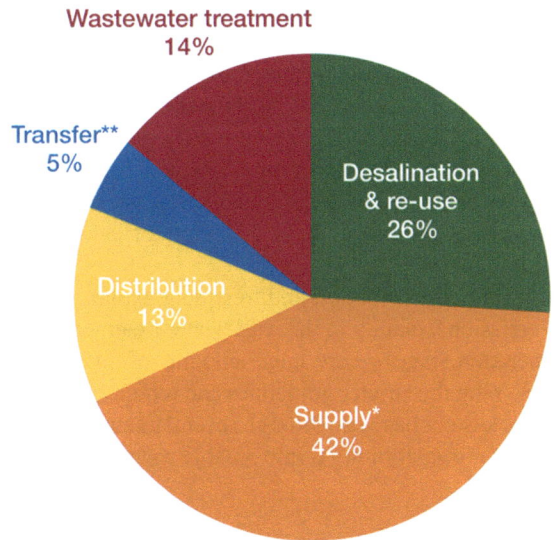

* Supply includes water extraction from groundwater & surface water as well as water treatment.
** Transfer refers to large-scale inter-basin transfer projects

Fig. 13.2 Global energy use in the water sector, 2016 (IEA 2018)

again. Membrane technology, also known as Reverse Osmosis (RO), is used in desalination to filter seawater by pushing it through RO permeable membranes, which inhibit the passage of dissolved salts. Operating the desalination plants requires high energy demands, and hence is a source of considerable greenhouse gas emissions, especially if the plant is operated by fossil fuels. RO desalination is until now the most advanced and competitive technology, as it is the most energy effective and relies entirely on electricity, and therefore could be run by renewable energy sources. This explains the dominance of RO plants, accounting for 84% of the total number of operational plants (Jones et al. 2019).

The conveyance and distribution of drinking water from seaside to cities, supply for agricultural purposes or water for industrial use would need to pump large volumes of water over large distances and elevations. This requires a considerable amount of energy, especially when it comes to bridging the height difference. As depicted in Fig. 13.2, distribution accounts for 13% of the total electrical energy consumption, most of which is generated by pump operations.

A further lever for energy savings requires the minimization of water losses along the distribution systems, caused by network leakages and pipeline bursts. Water pressure in the network together with aging infrastructure are the main challenges to utilities. Efforts to tackle these problems reduces the amount of potable water produced and treated, while still satisfying customer demands. Hence, the management of substantial leaks while pumping water through the distribution networks not only results in economic financial benefits, but also has an important environmental impact by saving water and reducing energy consumption and therefore greenhouse gas emissions.

Agriculture is globally the largest consumer of water (70% of freshwater in most regions (Scheierling and Tréguer 2018)). However, the sector is characterized by highly inefficient irrigation and drainage (I&D) systems, resulting in only a fraction serving its purpose for growing crops. The rest is drained, lost through unproductive evaporation or recharges ground and surface-water. Thus, improved water management practices are essential in the effort of achieving the sector's sustainability.

The wastewater collection, distribution and treatment explain an important part of the sector's GHG emissions, mainly in the wastewater treatment plants and especially in the sludge aeration stage, where huge amounts of air are blown into the sludge by compressors. With the stricter guidelines and requirements on the effluent discharge standards, black water has to go through several treatment processes before discharging it into nature or transforming it into potable water that could be fed back to distribution networks.

13.2.2 Opportunities for Carbon Emissions Reduction

AI and Big Data in the Water Sector

Artificial intelligence and machine learning are powerful tools to bring better efficiency and resilience to the water sector. Technologies are increasingly enabling device connectivity and various types of data to be captured in real-time (e.g. infrastructure, geographical information, consumption, water quality, etc.), which makes it possible to manage the water network proactively and intelligently, by achieving transparency, providing valuable insights, as well as translating information into knowledge and actionable measures.

Data mining and analytics techniques come into play while monitoring the demand trends and consumption patterns to generate forecasts, which serve operation decision making and allow optimal system planning and management. Additionally, the real-time monitoring of infrastructure (e.g. pipes, valves, pumps, reservoirs, tanks, junctions) makes it possible to foresee downtimes and predict faults and anomalies thanks to pattern recognition techniques and thus enable preventive measures and predictive maintenance strategies.

Digital twins and model-based simulations are becoming increasingly important to continuously optimize plant design and operating conditions, automate routine processes, support decision making and streamline maintenance by aiding repair and service.

The strength of the AI systems lies in their self-learning characteristic in a sense that they can continuously learn from the computed results, in order to finetune their recommendations and improve the quality of its output. By automating routine processes, detecting and mitigating problems at an early stage, maintaining asset health and optimizing the use of water and of the infrastructure, it is possible to reduce electricity consumption resulting in a decrease in emissions from the water sector.

Desalination Process Optimization and Renewable Energy Operation

The WEO report highlighted the impact of desalination treatment plants on the sector's energy consumption. In order to be able to answer the increasing consumer demand, desalination projects are expected to double the energy consumption by 2040 (IEA 2016). Regions like the Middle East and North Africa rely on desalination to satisfy the increasing demand. Reverse osmosis (RO) desalination plants are replacing the energy-intensive thermal distillation processes today, as they are more energy-efficient and have a lower unit production cost.

The increasing number of RO plants still contribute to the rising energy needs, due to their reliance on electric pumps to defuse water through the membrane at high pressures. In fact, operated with fossil fuels, their carbon footprint cannot be ignored. Hence, an alternative to reducing carbon emissions consists in switching to more sustainable energy sources, e.g. large-scale solar-powered water desalination plants (Siemens 2019). As the implementation of renewable energy sources is not a water industry-specific topic, it will not be the focus of this study and therefore will not be taken into consideration in our estimates, although it can be a significant lever for a green water sector. In addition to that, and despite the technological maturity of RO plants, digital twins and AI technologies still can provide operational intelligence by optimizing plant operations based on plant performance simulations relying on data and model-based analytics. Depending on the existing plant efficiency, savings between 0.5 and 5% (1–3% on average) in energy consumption can be achieved.

Pump Efficiency and Network Optimization

While pumps have reached mature technological advancements in terms of efficiency and optimized energy consumption, there is a potential for energy savings through the optimization of the whole network operation and definition of pump schedules that allow the pumps to work within their best efficiency range. Simulation models help determine the optimal structures or the optimal operating conditions of the distribution systems taking into consideration several parameters and constraints like demand forecasts, pipes and valves hydraulic characteristics and geographic distribution of consumers, as well as making better use of gravity in the network planning. The goal is to reach operational efficiencies reducing both wastage and shortage, i.e. delivering the adequate quantity with the required pressure in the right time to the consumer. Hitherto, a real-time control system defines and updates pump schedules by optimizing the water allocation based on the distribution network model and input from SCADA systems and demand forecast of DMAs.

Based on our operational experience, 2–5% pump energy savings can be achieved by the optimization of variable-speed drives, which account for about 70% of pumping energy in distribution networks. For the remaining 30%, a better pump schedule planning and use of pressurized water from elevated storage tanks can save 5–8% of energy.

Water Loss Reduction in Distribution Networks

Water production (e.g. desalination, treatment) and distribution are the largest consumers of energy, which would be wasted in case water does not reach the end consumer. This is mainly caused by leakages and bursts (while missing or faulty metering systems have solely a commercial aspect). It is one of the lowest hanging fruits that could be controlled through efficient strategies and that would improve the efficiency of water networks around the world. Although, it is difficult to estimate the global water losses, some studies suggest that the average is between 25 and 35% and that it can be even higher in developing countries lying in a 50–70% range (Hvilshoj 2015, FLUENCE NEWS TEAM 2020, AVK Group 2017). A study from the International Water Association (IWA) estimates the global volume of water losses to be 126 billion cubic meters of the water supplied per year generating approximately costs of USD 39 billion dollars a year (Liemberger and Wyatt 2018).

In many cases, water utilities already have a large amount of flow, level and pressure data available, which can be assessed in order to create transparency about the overall water balance and identify areas for action. This simple approach does not require large investments in additional sensors. Therefore, the detection of pipeline leakages and bursts becomes an easy task and the focus can then lie on the repair of the affected pipe segments.

By dividing the network into District Metering Areas (DMAs) and setting up meters for the consumers in these areas, data can be gathered and advanced systems for detection and management of pipe breaks and leaks can be implemented. These systems rely on AI-based algorithms to predict these events leveraging the available data, such as real-time sensor input, well-defined DMAs, SCADA and GIS data.

A hydraulic model, which replicates the nonlinear dynamics of the network and reproduces the real transportation and distribution behavior, can be established and calibrated, allowing the simulation of different scenarios, e.g. pipe isolation, closing valves, changing asset settings to mitigate a leak damage. The results can help identify critical parts of the system, reduce pre-location time of any leak or bursts, as well as optimize pressure control for sustainable leakage reduction. On the one hand, an early detection of leaks helps utilities avoid secondary damages and repair the pipes before major breaks. This will result in less water pumping and treatment, hence maintain the health of water facilities and save maintenance and operation costs. On the other hand, smart pressure management can avoid pipe stress and subsequent damages and will reduce considerably the water losses through leakages and bursts, especially during the lower network load.

The continuous leak management, early detection and localization, as well as timely repair measures—considering the economic aspects—can reduce the leakage rate by roughly 10% points, and even more in developing countries.

Improvement Wastewater Treatment Process and Management

In order to comply with the high discharge and reuse standards, the energy consumption associated with wastewater treatment plants (WWTPs) is becoming increasingly higher. It is therefore important to adopt efficient energy management measures in this stage of the water cycle and with it reduce the carbon emissions. Technological advances have made it possible to recover the by-products (such as energy, nutrients and biogas) generated throughout the treatment process within the plant. The energy can be exploited for cooling or heating or even for generating electricity. Thus, the plant can reach energy neutrality and reduce its carbon footprint to nearly zero. This is only realistic for large central sewage treatment plants, which account for about 20% of the global capacity (Renewable Energy Focus Journal 2015).

The automated operation of sewage treatment plants offers a great potential for data acquisition and evaluation for process optimization. It is necessary to make use of the IoT-technologies for remote monitoring and governance. The implementation of advanced automatic control systems taking into consideration the physical/chemical/biological interactions between different variables, measurable disturbance effects and key energy consumption indicators for monitoring and automatic operations. In a sludge treatment process, the aeration of biological purification steps represents the largest energy consumer, as it accounts for 60% of the total energy consumption (Guo et al. 2019). A model predictive control (MPC) could adjust the aeration in time maintaining the required conditions for nitrification by preventing excessive aeration, reducing the negative effects of influent fluctuation and as result achieve energy efficiency.

Process optimization for wastewater treatment plants can achieve roughly 3–5% energy savings. For the largest plants (responsible for 20% of the global sewage treatment capacity), holistic management can incur significant savings (20–50%).

13.3 Conclusions

The water sector is a non-negligible contributor of global GHG emissions. By leveraging the energy-water nexus, savings in water and energy as well as a greener industry can be achieved in light of the recommendations presented in this chapter:

(1) Water loss reduction in the pipelines is an ideal first step towards simultaneously improving water and energy efficiency.

(2) The adoption of a wide range of existing and developing technologies (e.g. big data, modelling and simulation, data analytics) for process optimization in desalination plants and wastewater treatment plants has the potential to reduce energy consumption and therefore achieve appreciable emissions reductions.

(3) A holistic network optimization and efficient pump operation are software-based solutions that do not require much preliminary investment.

Table 13.1 shows a summary of the estimated energy saving potentials based on actual energy consumption figures. With the implementation of existing technologies and solutions, energy savings between 10 and 16% could be reached. Further investments in R&D on low carbon solutions could result in even greater energy savings.

It is important to mention that an additional approach to make these systems energetically sustainable is by introducing renewable energy sources (power grid). Solar energy has to be further adopted in WWTP and desalination plants.

While leakage management is the biggest lever, followed by the optimization of large treatment plants, both require substantial investment in infrastructure. Indeed, the main hurdle in implementing these types of solutions is their implementation costs. As a matter of example, detection of leaks is fairly simple from a technical perspective, but the repair is costly, and the available limited resources must be prioritized. Water utilities would have to plan their repair activities, accordingly, based on the level of leakage and an economic assessment of the available measures (Hendrickson and Horvath 2014).

Based on the global energy consumption of the water sector and a saving potential of 10–16%, a realistic energy reduction target would be between 11.5 and 19.5 Mtoe.

13.4 Outlook

The growing world population leads to an ever greater demand for water, so that the proposed savings are likely to be quickly overweighed. This makes it increasingly important to use water as sparingly as possible and to produce it in an energy-efficient manner. This article, which is essentially based on the authors' collected experience, is intended to provide food for thought on this issue and a lead-in to the discussion on how to improve climate resilience by reducing demand on scarce water resources.

We have been able to demonstrate examples, where digitalization is available already today and ready for implementation. Most of the proposed solutions can be

Table 13.1 Breakdown of potential energy savings in the water industry through implementation of existing technologies

Reference values							
Total energy consumption in the water sector	120 Mtoe						
Share of desalination	26%						
Share of supply	42%						
Share of distribution	13%						
Share of transfer	5%						
Share of wastewater treatment	14%						

				Worst case estimation		Best case estimation	
Levers for water, energy and cost savings	Action variable	Share of total energy consumption (%)	Amount of energy consumed (Mtoe)	Estimated savings (%)	Smart water/energy savings (Mtoe)	Estimated savings (%)	Smart water/energy savings (Mtoe)
Water production							
Desalination process optimization	Energy	26	31.2	0.50	0.16	5	1.56
Water distribution							
VSD operation optimization	Energy (70%)	9	10.9	2	0.22	5	0.55
Pump schedule optimization	Energy (30%)	4	4.7	5	0.23	8	0.37
Desalination, water supply and distribution							

(continued)

Table 13.1 (continued)

Levers for water, energy and cost savings	Action variable	Share of total energy consumption (%)	Amount of energy consumed (Mtoe)	Worst case estimation		Best case estimation	
				Estimated savings (%)	Smart water/energy savings (Mtoe)	Estimated savings (%)	Smart water/energy savings (Mtoe)
Management of physical water losses (Leakages, bursts, etc.)	Energy	81	97.2	10	9.72	15	14.58
Wastewater treatment							
Process optimization	Energy	14	16.8	3	0.5	5	0.84
Holistic plant management toward energy self-sustainability for large plants	Energy	3	3.4	20	0.67	50	1.68
Totals for the reference value 120 Mtoe				10	11.5	16	19.58

applied for existing plants, some of them based mostly on existing sensor information with only little CAPEX spending for additional hardware. Therefore, SMART CITY initiatives should also consider the water nexus as an important area for the reduction of GHG emissions with a substantial potential.

References

AVK Group (2020) AVK non-revenue water solutions 2017. Accessed 2020

FLUENCE NEWS TEAM (2020) What is non-revenue water? Fluence, 6 November 2019. https://www.fluencecorp.com/what-is-non-revenue-water/. Accessed 22 July 2020

Guo Z, Sun Y, Pan S-Y, Chiang P-C (2019) Integration of green energy and advanced energy-efficient technologies for municipal wastewater treatment plants. Environ Res Public Health 16(7):1282

Hendrickson TP, Horvath A (2014) A perspective on cost-effectiveness of greenhouse gas reduction solutions in water distribution systems. Environ Res Lett 9(2):021017

Hvilshoj S (2015) Reduction of non-revenue water around the world.(IWA) International Water Association, 20 May 2015. https://iwa-network.org/reduction-of-non-revenue-water-around-the-world/. Accessed 22 July 2020

(IEA), International Energy Agency (2017) Water energy nexus—excerpt from the world energy outlook 2016. IEA Publications, Paris, France

(IEA) International Energy Agency—World Energy Outlook 2018 Excerpt (2018) Energy, water & the sustainable development goals. IEA, Paris

Jones E, Qadir M, Vliet MTV, Smakhtin V, Kang S-M (2019) The state of desalination and brine production: a global outlook. Sci Total Environ 657(03):1343–1356

Liemberger R, Wyatt A (2018) Quantifying the global non-revenue water problem. Water Sci Technol Water Supply 19(3):ws2018129

Renewable Energy Focus Journal (2020) Siemens helps transform the main wastewater treatment plant in Vienna into a green power plant. Elsevier Ltd, 30 November 2015. http://www.renewableenergyfocus.com/view/43334/siemens-helps-transform-the-main-wastewater-treatment-plant-in-vienna-into-a-green-power-plant/. Accessed 07 July 2020

Scheierling SM, Tréguer DO (2018) Beyond crop per drop—assessing agricultural water productivity and efficiency in a maturing water economy. international development in focus. World Bank, Washington, DC

Siemens AG (2020) Siemens equips desalination plants in Saudi Arabia with process automation. Smart Water Magazine, 21 October 2019. https://press.siemens.com/global/en/pressrelease/siemens-equips-desalination-plants-saudi-arabia-process-automation. Accessed 07 July 2020

The World Bank Group—Water Global Practice (2019) The role of desalination in an increasingly water-scarce world. The World Bank Group, Washington DC

(WaCCliM), Water and Wastewater Companies for Climate Mitigation (2014) Linking water and climate—greenhouse gas reductions in the water sector. WaCCliM, Eschborn

Chapter 14
Insights: Utilities Decarbonisation

Christoph Dörr

1. What impact does the government's promise to reduce CO_2 emissions have on the operations of Stadtwerke Pirmasens?

By 2030, the German government aims to meet its climate target with a reduction of at least 55% in greenhouse gas emissions compared with 1990 levels, and to achieve greenhouse gas neutrality by 2050. To achieve this, CO_2 emissions will be priced on gasoline, diesel, heating oil and gas. This price will increase over the years.

As a municipal utility, we support the government's climate targets. Although medium-sized, we are innovative in developing environmentally friendly technologies. To implement our ideas, we looked for local partners with whom we founded the Winzeln Energy Park. Today, biogas is produced there from green waste. We feed this into our natural gas network with the help of a biogas transfer station, which raises the quality of the biogas to that of natural gas. The Winzeln Energy Park also serves us as a development laboratory for further innovations. In addition, we are investing in the modernisation of our infrastructure because we are convinced that modern networks can, on the one hand, better adapt to population development and, on the other hand—in their function as storage facilities—make a significant contribution to reducing CO_2 emissions. But we are also starting with ourselves. Back in 2015, we established an energy management system which led to a significant increase in energy efficiency in our buildings and plants. In addition, our investments have been profitable. Many public institutions in our city benefit from our good experience by supporting them in carrying out energy efficiency audits.

2. What role does digitalisation play in the municipal sector today?

I see this as a great opportunity. There is practically no area of life that can currently escape digitalisation. On the contrary, anyone who lags behind in this development

C. Dörr (✉)
CEO, Stadtwerke Pirmasens, Pirmasens, Germany
e-mail: Doerr.C@Stadtwerke-Pirmasens.de

© Springer Nature Switzerland AG 2022
O. Inderwildi and M. Kraft (eds.), *Intelligent Decarbonisation*, Lecture Notes in Energy 86, https://doi.org/10.1007/978-3-030-86215-2_14

will soon be left behind and no longer competitive. In private life, robots are increasingly establishing themselves as household helpers. Many people wear so-called "wearables"—mobile data processing devices such as the smartwatch—and can use them not only to monitor fitness but also to transmit health data directly to the doctor. In the municipalities, there are longer decision-making paths, which also prolong the implementation of projects. This is precisely where digitalisation can help speed up these processes. For the municipal sector, the question is therefore not whether the topic will be taken up, but to what extent municipalities want to benefit from digitalisation.

3. Do you see any potential in digitalisation for reducing the city of Pirmasens' CO_2 emissions?

I take an ambivalent view of this: for some, digitalisation will contribute to a massive increase in environmental problems; for others, digitalisation is the decisive key to solving precisely these problems. The crucial question is therefore whether, at the end of the day, digitalisation helps to avoid at least as much CO_2 emission as it causes. In the long term, digitalisation can lead to a reduction in CO_2 emissions, as developments continue to advance. Currently, however, it tends to be a driver of emissions. According to a study by the French think tank The Shift Project, the digital industry already produces almost four percent of global emissions, around twice as much as the entire air traffic sector. And the trend is rising sharply. If growth continues unabated, it is foreseeable that this figure will double by 2025.

At the same time, it is undisputed that digitalisation is one of the key factors for a successful energy transition. Autonomous smart grids, intelligent load management, forecasting models and green energy based on digital tokens: the energy transition is also a digital transition. So, as with all technologies, there are advantages and disadvantages. In order to compensate for the disadvantages that still exist in the future, we need appropriate innovations. Digitalisation must not take place at any price and must be ecologically sensible.

4. If there were one problem facing you as the CEO of Stadtwerke Pirmasens that you could solve with the use of AI, what would it be?

AI is not a general problem solver. However, for energy supply in particular, we can use AI for the energy transition and climate protection. We can't avoid dealing with AI because our future competitiveness is at stake. For example, analysing environmental data and its progression enables us to predict emissions. We are also already developing innovative applications to optimise heat generation using AI systems to reduce greenhouse gas emissions, save resources and lower generation costs. For us as a municipal utility, I would hope that the use of the "right" algorithms would lead to intelligent networking, control of renewable energies and energy storage systems, which would optimise energy supply—including mobility—in a smart grid.

Christoph Dörr is the CEO of Stadtwerke Pirmasens, the municipal utility and service company of the city of Pirmasens, Germany. Prior to taking up this position, Christoph was the Sole Director and then Managing Director of Stadtwerke in Bad Oeynhausen, North Rhine-Westphalia for a period of 12 years. He received his degree in Economics from Saarland University. Since joining Stadtwerke in Pirmasens, Christoph has led the organisation in developing innovative digital technology for energy efficiency and CO_2 emissions reduction.

Chapter 15
Decarbonisation of the Urban Landscape: Integration and Optimization of Energy Systems

Salvador Acha, Edward O'Dwyer, Indranil Pan, and Nilay Shah

Abstract We highlight the key pillars of urban energy systems which would leverage on AI and digital technologies for a low carbon future. We summarise a couple of real world applications where optimisation, intelligent control systems and cloud-based infrastructure have played a transformative role in improving system performance, cost-effectiveness and decarbonisation. The case studies show that AI and digital technologies can be implemented for standalone unit operations to achieve such benefits. However, more importantly as the second case study shows, applying such technologies at a system level by integrating multiple energy vectors would give much more flexibility in terms of operation, resulting in better performance improvements and decarbonisation strategies. We conclude by highlighting the strategic trends in this fast evolving field and giving a broad outlook in terms of cost reductions and emissions savings for similar intelligent energy systems.

15.1 Introduction

With high energy consumption across different sectors and little space for renewable generation, the urban setting poses many barriers to decarbonisation. Nonetheless, the emergence of digital technologies and deployable computational intelligence provides cause for optimism (Rolnick et al. 2019), signalling a transition from centralised energy suppliers and fragmented demands towards more distributed suppliers and coordinated multi-vector demands. Such a transition would allow for low and zero-carbon sources to play a more dominant role, particularly encouraging the onset of greater renewable generation by enabling the intelligent shifting of demands to match

S. Acha · E. O'Dwyer · N. Shah (✉)
Centre for Process Systems Engineering and Department of Chemical Engineering, Imperial College London, London SW7 2AZ, UK
e-mail: n.shah@imperial.ac.uk

I. Pan
Centre for Process Systems Engineering and Centre for Environmental Policy, Imperial College London, London SW7 2AZ, UK

The Alan Turing Institute, The British Library, London NW1 2DB, UK

© Springer Nature Switzerland AG 2022
O. Inderwildi and M. Kraft (eds.), *Intelligent Decarbonisation*, Lecture Notes in Energy 86, https://doi.org/10.1007/978-3-030-86215-2_15

the supply needs. The requirements for successful deployment of digital technology in this context can be characterised by four pillars:

- Sensing and data fusion;
- Prediction;
- Decision support;
- Actuation.

Increased sensing capabilities facilitate the acquisition of more and better-quality data, prediction allows for this data to be processed into useful information, decision support leverages optimization and related techniques to turn this information into tangible decisions which are then acted on with appropriate actuation strategies.

The role of these key pillars in the decarbonisation process is explored in this chapter through the illustration of several activities undertaken in the UK with the involvement of the authors. A case study is first presented in which these stages are deployed in the control of a CHP system in a commercial building to enable more intelligent operation. The focus then moves to district energy schemes, where the use of modelling and optimisation techniques can result in significantly improved design choices, while overcoming challenges in understanding the complex interactions between different energy sectors. Following this, the application of predictive control to enable the integration of electrified heating and transport is discussed through interventions underway by the local government in London to establish heat-pump based communal heating and increased levels of electric vehicle uptake without compromising the pre-existing grid infrastructure. Such an approach combines sensing, prediction and decision support, ultimately leading to actuation that facilitates significant decarbonisation.

15.2 Computational Intelligence in Smart Cities

15.2.1 Overview of Key Elements

The four key pillars, as mentioned before, leverage on advances in multiple fields in the last couple of decades and their symbiotic integration has the potential to lead towards transformative digitalisation and a low carbon future. With regards to the first pillar of sensing and data fusion, the very low cost of sensing electronics and IoT (Internet of Things) devices, coupled with high bandwidth internet connections and cheap cloud storage, have made it possible to gather a lot of data about the system and also store and process such large datasets. Data processing and fusion techniques have been accelerated both due to edge computing in smart sensors as well as low cloud computing costs. Similarly prediction algorithms (second pillar) based on pure data-driven approaches have seen a resurgence in this decade due to breakthroughs in deep learning and allied machine learning techniques like generative adversarial networks. Although these techniques have gained a lot of traction for unstructured datasets like images and text, their 'black box' nature has come under increased scrutiny for engineering applications where safety and reliability are major concerns. The need for machine learning in such engineering applications has led to researchers adopting rule-based or some hybrid learning techniques leveraging both symbolic logic and deep learning. Probabilistic frameworks in machine learning (which can provide uncertainty estimates with the predictions) have gained increasing traction due to cheap computation which has been able to unlock hitherto intractable numerical approximations of generic Bayesian statistical models. They also allow for more interpretable models and have led to the development of multiple probabilistic programming languages to facilitate such model building, validation and comparison. Such interpretable probabilistic frameworks are key drivers for effective decision support systems (pillar three) in the context of smart energy systems. The final aspect of actuation (pillar four), basically closes the loop by taking real-time actions on control and coordination of such smart energy networks. These closed-loop systems facilitate the data-driven analysis to undertake continuous learning, decision making and adaptive update of control strategies. The following sections delve into more practical detail by reviewing practical case studies based on this four-pillar framework.

15.2.2 Real-Time Operation Control of a Combined Heat and Power (CHP): A Case Study

Co-authors of this chapter have been involved in an innovative trial testing the capabilities intelligent control systems can bring to an operating CHP system in a UK supermarket (Olympios et al. 2020). This research in collaboration with an industrial partner has allowed us to quantify the value these decision-support tools can have in improving the performance of a distributed generation system found across energy-intensive buildings or districts. Smart control system technologies for CHP systems are scarce in the literature and have the potential to generate significant savings. Research suggests minimal capital investment is required in hardware and software development. The live cloud-based solution is presented next covering its methodology and results both from the test-bed site and from digital twins developed for other sites where similar controllers could be installed.

15.2.2.1 Background

Although extensive research has been performed on theoretically optimizing the design, sizing and operation of CHP systems, less effort has been devoted to understanding the practical challenges and the effects of uncertainty in implementing advanced algorithms in real-world applications. Most CHP control systems already allow 'load-following', matching output to the store's electricity demand, however in this situation the variation in electricity import and export prices is not fully exploited. As electricity markets evolve and introduce dynamic tariffs there will be a growing opportunity for energy services that can intelligently control supply and demand for financial benefit. This case study implements a real-time control approach for CHP optimal operation in commercial buildings to maximise the actuation impact high-speed computing can have on physical systems as a result of the price signals it receives.

15.2.2.2 Methodology

The proposed CHP optimizer aims at making the best possible operation choice by continuous monitoring and automation via a two-level control system consisting of a cloud-based controller and an onsite controller. The function of the cloud-based controller is to utilize continuous real-time measurements to solve the optimization problem and therefore plan the energy strategy of the system. The onsite controller then actuates the instructions received by the cloud-based controller, acting as a safety system and secondary control logic in case of any wrong instructions or unexpected problems arise. A schematic representation of the environment in which the CHP engine operates under is provided in Fig. 15.1. The inputs required are energy price forecasts and the heat and electricity demand of the facility. This setup allows decision-makers to operate the CHP system in a continuous manner according to their preferred strategy or objective function (e.g. costs). The optimizer allows electricity price forecasts to be considered in the CHP unit operation. Exported energy can be maximised during high export tariffs and similarly imported energy can be minimised when favourable. The cloud-based controller works through the Amazon Web Service (AWS), utilising its servers to provide substantial cost reduction over traditional non-cloud-based methods. The software communicates with the on-site controller, setting the part load output for the engine, with the on-site controller subsequently maintaining the unit at that point.

Prior to the part load optimal choice, the optimization problem requires several inputs. Firstly, it requires the engine model coefficients, which provide relationships for the fuel input, power output, and thermal output of the CHP engine as a function of part load output. These values are all fixed and are stored in the cloud-based controller. Then, at each time step, real-time data are obtained for the site electricity and heat demand, CHP engine and grid carbon factors, electricity import prices, as well as forecasts of exported electricity prices and the prediction intervals around them. Once the optimization problem is solved, the engine's operation is set at the chosen part load and the costs associated with the operation choice are registered. The process is repeated whenever data are published from the onsite controller to the cloud-based controller.

Fig. 15.1 Schematic representation of the intelligent CHP control system (Olympios et al. 2020)

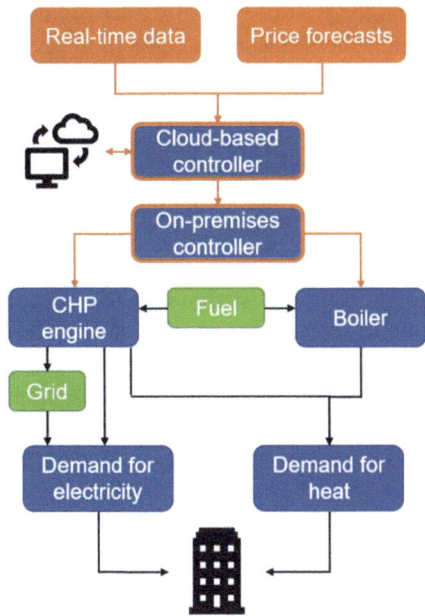

15.2.2.3 Results

Both the on-site optimizer and its digital twin model were validated within acceptable error margins. Below, we summarise the results obtained.

The cloud-based CHP control optimizer technology was installed in an existing 250 kWe CHP system within a mid-size supermarket in the UK. In addition, a simulation (i.e. digital twin) of the optimizer was also conducted in 2 other stores with CHP systems with the purpose of assessing the controller benefits prior to embarking on a series of investments rolling out the solution. Results indicate the electrical load of the site with respect to the full electrical output capacity of the CHP unit has a major role in the benefits the real-time CHP controller provides; this is primarily due to the high cost of the electricity commodity. The annual average daytime load (ADL) defined as the store's electricity consumption divided by CHP maximum output helps us understand the benefits of the optimizer.

For the test-bed site with a 43% ADL, which we shall call 'Store A', the financial benefit is marginal, only a 2% improvement against the baseline CHP strategy of solely being on and operating at a fixed output during store trading hours. By operating in a dynamic manner, savings are generated either via exporting during periods of high export price, mainly during the afternoon, or operating the CHP overnight, when the baseline control system has the unit off. Since electricity import prices are significantly higher than the export prices, it makes avoiding electricity imports particularly lucrative. It is unfortunate that for the 'live' case study the CHP system is oversized and thus the capabilities of quickly ramping up/down the engine are not fully exploited because the baseline operation is sufficient for this particular site. Nonetheless, it is worth highlighting that the hardware costs of the installation is minimal, circa $10,000 USD. Meanwhile, software development costs were estimated at $200,000 USD, although a significant figure, the coding architecture is robust and easy to modify; thus reducing further replication and updating costs. Although, the trial was a technical success (with no glitches nor unattended consequences on engine performance), due to the poor financial benefit obtained the cloud-based controller platform was tested through simulations via a digital twin in other CHP sites of the retail company.

The digital twin simulations were conducted in 'Store B' with a 400 kWe CHP system and an ADL of 75% and in 'Store C' with a 210 kWe CHP system with an ADL of 138%. Results for 'Store B' suggest a financial benefit in the order of 12% as the generator has the freedom to maximise its operational flexibility; indicating that the generator should be running even when the store is closed as it is economically advantageous. The payback period for installing the real-time controller in 'Store B' is less than 2 years. Meanwhile, in 'Store C' the financial benefit is in the order of 7%. In this site, an identical CHP operation as with the baseline control system was identified, with relative savings generated from overnight operation. The payback period for installing the real-time controller in 'Store C' is less than 3 years.

Lastly, it is worth remarking that despite the technological prowess of controlling a CHP system in real-time, the 'smarter' operation of the engine to maximise financial returns has its caveats. The most important one is that due to the nature of the heat load profiles in supermarkets, a considerable amount of excess heat is underutilised; particularly for sites where the ADLs are higher. Unless CHPs are coupled to local district heat networks, this resource will be wasted. On the other hand, the environmental impact smarter operations can have is detrimental if the fuel of choice is natural gas. Maximising the use of a CHP system increased the carbon emissions for 'Store B and C' by 35 and 40%; respectively. There is an imperative need to power these CHP systems with low carbon sourced fuels, otherwise the carbon mitigation from CHP use will be marginal and in some cases worse than utilising electricity from the grid (Acha et al. 2018).

15.3 The Multi-vector Design and Operation Challenge: Optimising the Low Carbon Transition

Understanding the benefits and limitations new distributed technologies can have in the existing built environment is paramount to comprehend under what conditions it is best to exploit the virtues integrated energy systems offer. In this section we highlight the insights obtained from modelling distributed energy cyber-physical systems for specific built environment case studies in the UK. These problems focus on addressing design and operational objectives that arise from considering multi-vector energy streams as a result of a wide portfolio of distributed energy systems available to end-users today that support district heat network (DHN) schemes.

15.3.1 District Energy Systems Examples in London, UK

Recent research conducted in collaboration with local authorities in London, UK suggests district energy schemes will play a pivotal role in urban environments guaranteeing the security of supply and sustainability in electricity and heat provision (Greater London Authority 2018a). Although authorities and policy makers are convinced district energy schemes are valuable, they are concerned with quantifying the added value they offer by expanding their customer outreach; particularly with regards to supplying heat to a greater number of end-users and supplying electricity to a new customer base such as electric vehicle (EV) fleets.

Delangle et al. (2017) explored the optimal expansion of a district heating network in the Isle of Dogs area in South East London. Through a MILP approach a mix of technologies was evaluated to quantify the preferred outcomes if the scheme optimised its expansion design for either cost or environmental minimisation. The existing network chosen supplies heat to 22 buildings using 2.5 km of pipes. In the existing energy centre, two CHP plants, two natural gas boilers and two thermal stores ensure the continuous supply of heat to the connected dwellings. The extension considered for this network includes 31 buildings pre-identified, their connection requiring the installation of 3.5 km of new pipes. Results showed the preferred approach was to consider an incremental connection of buildings, clusters with a high heat demand being connected first followed by the small clusters with a low heat demand. This solution suggests that over a 12-year period the cost balance against a 'do-nothing' approach exceeds £25 million while also achieving a carbon footprint that is reduced by two-thirds thanks to the installation of biomass boiler systems and an increased use of thermal storage facilities. Due to the energy market conditions, CHP electricity production is championed resulting in attractive revenues from the export of energy to the local distribution network.

The economic drivers to self-generate electricity in district schemes and not only operate to satisfy heating requirements were researched in detail for the borough of Islington by Chakrabarti et al. (2019). The premise of this work proposes the district energy scheme to not sell electricity to the local operator but instead try to agree supply contracts with commercial EV fleets. Using the electricity generated on-site to power EVs can make district heating networks more economically feasible, while also increasing environmental benefits. The approach here combines agent-based models to represent EVs and their 'temporal-spatial' loads are solved by a MILP that has the goal to maximise revenues. The Borough of Islington has a district heating network that serves 850 homes and two leisure centres, called the Bunhill Heat and Power network. Bunhill I is the current energy centre where a 2.4 MWe natural gas CHP engine (CHP 1) is used to produce heat and electricity that is supplied to the council. This has led to reductions of around 10% in the heating bills for consumers benefited by the scheme, in addition to a reduction of up to 60% in CO_2 emissions. There is also a 115 m^3 thermal store, which may be used to store heat for times of known high heat demand. A new energy centre (Bunhill II) is being built to supply heat to at least 454 additional homes, a school and a nursery as an extension to the existing DHN. In addition to an extra CHP unit (CHP 2), Bunhill II will have a heat pump (HP) that uses electricity supplied by CHP 2 to produce heat. This is an air-source heat pump that can take heat from the 20–30 °C London Underground Limited (LUL) ventilation air and use it to efficiently heat up water to around 80 °C. A key aspect of this analysis focuses on understanding the energy market dynamics of sourcing fuel and electricity as well as re-selling electricity. Results indicate that as EV adoption increases the financial drivers to fully exploit CHP electricity production is preferred increasing the profitability of district schemes. For instance, it was found that a 30% adoption level of EVs increases revenues by about 11% and reduces local emissions marginally.

15.3.2 Environmental Impact from Low Carbon Fuels

While both DHN case studies discussed above clearly indicate there are considerable financial gains to be made by optimising the design and operation of district energy systems, the environmental benefits are much harder to estimate as they are subject to the very peculiar details of each project and the mix of technologies considered. Nonetheless, it can be argued that if biomethane can be sourced instead of natural gas to run CHP systems in the UK, the environmental benefit can be in excess of 90% as discussed by Cedillos Alvarado et al. (2016) when conducting optimization modelling of cogeneration schemes in commercial buildings. The environmental benefits from what appears to be more benign fuels at the end-use point can be considerable, however they need to be carefully analysed since the carbon emissions can be displaced to other parts of the value-chain be it for biomethane, hydrogen or any other fuel that at a first glance appears to be environmentally friendly (Langshaw et al. 2020).

15.4 Integration of Energy Systems

The traditional reliance of heating and transport sectors on fossil fuels has largely precluded the need for integrating the various decentralised subsystems involved. The decarbonization of the power grid has provided an incentive for electrification of these sectors. However, an energy landscape dominated by electrical demands and relatively inflexible renewable sources poses many challenges, as previously disparate systems become operationally interconnected. Coordinated energy management is paramount to ensure system operational constraints and security of supply requirements are adhered to while low-carbon resources are maximised. Faced with these challenges, an urban setting provides unique opportunities to make use of intelligent cyber-physical systems due to the presence of large collocated energy demands which, with appropriate use of advanced control and computational intelligence techniques, can be coordinated in a manner that reduces the overall burden on the power grid and/or other energy sources. As a result, many city-level strategies (e.g. the London plan for zero carbon (Greater London Authority 2018b) or Copenhagen's plan for carbon neutrality by 2025 (The City of Copenhagen Technical and Environmental Administration 2012)) focus not just on electrification, but also on the need for smart technologies to make such a transition possible.

15.4.1 District Level Coordination

While the need to shift flexible demands, utilise energy storage assets and maximise decentralised energy sources may be clear, doing so requires technologies for forecasting demands and production profiles, optimising operational decisions and providing secure communication between various subsystems. The demonstration of such technologies requires access to multiple large-scale energy systems and as such, many research activities in the area have focused on local government-managed systems. One such example is the Horizon 2020 funded Sharing Cities project (SHARING CITIES 2021). The London branch of Sharing Cities has seen the council of the Royal Borough of Greenwich undertake ambitious electrification activities including heat pump retrofit in their social housing stock, increased photo-voltaic (PV) solar generation and increased electric vehicle (EV) charging infrastructure (see Fig. 15.2). These different solutions are to be linked through the development of an intelligent energy management framework that combines AI-based forecasting services with Model Predictive Control (MPC) to optimally manage various energy subsystems while coordinating them at district level. Studies have been carried out to evaluate the impact of these interventions, indicating that a CO_2 emission reduction of about 77% could be achieved with operational cost savings of over 30% (the capital investment costs needed for these activities are not accounted for in these figures) (O'Dwyer et al. 2020). Furthermore, the intelligent management of this infrastructure allows for the shifting of EV load to maximise local PV utilisation (leading to about a 10% reduction in EV charger demand-related CO_2 emission) as well as shifting thermal or electrical load to avoid violation of district-level grid constraints.

15.4.2 The Built Environment

The potential to make use of the energy flexibility offered by buildings to aid the wider energy landscape has given rise to the concept of an "Active Building" which can dynamically interact in real-time with the energy networks to which it connects (these concepts underpin the goals of the Active Building Centre for example (The Active Building Centre 2021)). In addition to any storage and generation assets present, the building itself can be used as an energy store by managing the internal temperature without deviating outside pre-defined user comfort bounds. In an urban setting, such a perspective has mostly been confined to district heating systems. The

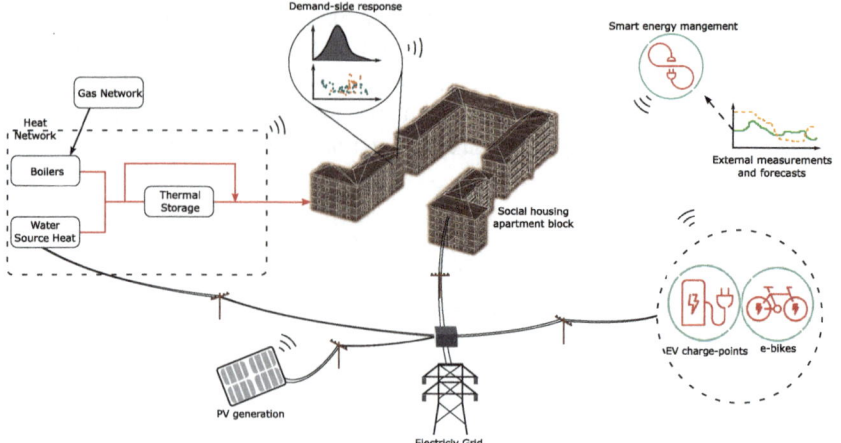

Fig. 15.2 Energy assets in Greenwich electrification activities

onset of electrification and the increase in renewable generation has greatly expanded the potential of such concepts. However, responsive buildings require intelligent control and communication strategies and MPC is a natural fit given the predictive capabilities and the presence of explicit constraints. The challenge of coordinating large numbers of decentralised buildings is tackled in Gonzato et al. (2019), whereby a price-coordination strategy was developed to ensure that high-level grid constraints could be adhered to by many individual heat-pump-supplied buildings without the need for computationally intensive centralised optimization. Such an approach not only allows for reduced infrastructure capacity, but also introduces the possibility for a grid operator to impose constraints in the system at times of grid stress. A simulated case study of 100 buildings showed that the strategy could be used to reduce electrical power peak demand by more than 50% without impacting user comfort. In itself, this technology does not reduce the carbon intensity of the built environment, however its deployment would enable a far greater uptake of low-carbon sources.

15.5 Conclusions

The confluence of new sensing technologies, new energy technologies, improved algorithms and distributed and cloud computing provides enormous opportunities for smart infrastructure in support of intelligent decarbonisation. Considering decarbonisation as a whole, it is both a technological and systems challenge. In particular, when thinking about the latter, it becomes clear that integration across infrastructures in both design and operation will become important to ensure cost-effective emissions reductions. Here we have described a four-pillar approach for intelligent design and operation. We particularly wish to emphasise the importance of the decision support and actuation pillars because there is a great deal of focus on sensing and data analytics already. Without better decision-making and implementation of the decisions, the potential benefits of new technology will be undermined.

15.6 Outlook

Our experience is that smart operation of existing systems can lead to significant cost and emissions savings; typically of the order of 10–30% reductions and with good payback periods. This is an excellent place to start because it quickly demonstrates the value of improved sensing, processing, decision-making and actuating methods. Further benefits will accrue when using these methods and the understanding of the wider system to implement design changes and greater integration across subsystems. This goes beyond pure operational optimisation and requires significant capital expenditure. In our experience, savings in resource consumption will offset some of the capital expenditure, but lower-carbon technologies and energy vectors will inevitably "cost" more than incumbent systems. However, we note that energy inputs are a small factor of gross value added (GVA) and our previous calculations demonstrate that deep decarbonisation of an economy, if done intelligently, can be performed at a cost of around 1.5–3% of GDP.

Acknowledgements This research was supported by funds provided via the Imperial—Sainsbury's Supermarkets Ltd. Partnership, UK. This work was also supported by the UK Engineering and Physical Sciences Research Council (EPSRC) [grant numbers EP/R045518/1] and by the EDF Energy R&D UK Centre as part of the SparkFund initiative. The Sharing Cities project received funding from the European Union's Horizon 2020 research and innovation programme under Grant Agreement No. 691895. The authors would also like to thank Siemens, the Greater London Authority and the Royal Borough of Greenwich with whom collaboration on the project has been carried out. IP would like to acknowledge funding from the Imperial College Research Fellowship.

References

Acha S, Mariaud A, Shah N, Markides CN (2018) Optimal design and operation of distributed low-carbon energy technologies in commercial buildings. Energy 142:578–591. https://doi.org/10.1016/j.energy.2017.10.066

Cedillos Alvarado D, Acha S, Shah N, Markides CN (2016) A technology selection and operation (TSO) optimisation model for distributed energy systems: mathematical formulation and case study. Appl Energy 180:491–503. https://doi.org/10.1016/j.apenergy.2016.08.013

Chakrabarti A, Proeglhoef R, Turu GB, Lambert R, Mariaud A, Acha S, Markides CN, Shah N (2019) Optimisation and analysis of system integration between electric vehicles and UK decentralised energy schemes. Energy 176:805–815. https://doi.org/10.1016/j.energy.2019.03.184

Delangle A, Lambert RS, Shah N, Acha S, Markides CN (2017) Modelling and optimising the marginal expansion of an existing district heating network. Energy 140:209–223. https://doi.org/10.1016/j.energy.2017.08.066

Gonzato S, Chimento J, O'Dwyer E, Bustos-Turu G, Acha S, Shah N (2019) Hierarchical price coordination of heat pumps in a building network controlled using model predictive control. Energy Build 202(109421). https://doi.org/10.1016/j.enbuild.2019.109421

Greater London Authority (2018a) London environment strategy. Technical report, London

Greater London Authority (2018b) Zero carbon London: a 1.5oC compatible plan. Technical report. December (2018). http://www.london.gov.uk/zero-carbon-plan

Langshaw L, Ainalis D, Acha S, Shah N, Stettler ME (2020) Environmental and economic analysis of liquefied natural gas (LNG) for heavy goods vehicles in the UK: a well-to-wheel and total cost of ownership evaluation. Energy Policy 137:111161. https://doi.org/10.1016/j.enpol.2019.111161

O'Dwyer E, Pan I, Charlesworth R, Butler S, Shah N (2020) Integration of an energy management tool and digital twin for coordination and control of multi-vector smart energy systems. Sustain Cities Soc 62(July):102412. https://doi.org/10.1016/j.scs.2020.102412

Olympios AV, Le Brun N, Acha S, Shah N, Markides CN (2020) Stochastic real-time operation control of a combined heat and power (CHP) system under uncertainty. Energy Convers Manag 216:112916. https://doi.org/10.1016/j.enconman.2020.112916

Rolnick D, Donti PL, Kaack LH, Kochanski K, Lacoste A Sankaran K, Ross AS, Milojevic-Dupont N, Jaques N, Waldman-Brown A, Luccioni A, Maharaj T, Sherwin ED, Mukkavilli SK, Kording KP, Gomes C, Ng AY, Hassabis D, Platt JC, Creutzig F, Chayes J, Bengio Y (2019) Tackling climate change with machine learning. http://arxiv.org/abs/1906.05433

SHARING CITIES (2021) http://www.sharingcities.eu/

The Active Building Centre (2021) Active building centre homepage. https://www.activebuildingcentre.com/

The City of Copenhagen Technical and Environmental Administration (2012) CPH 2025 climate plan. Technical report

Chapter 16
Insights: Infrastructure Management

Calvin Chung

1.   What impact has Singapore's plan to reduce CO_2 emissions on JTC?

JTC's sustainability efforts are aligned with Singapore's nationwide plans to tackle climate change and its target by 2050 to halve emissions from a planned 2030 peak. Sustainability is becoming a key concern for investors and manufacturers in response to consumers' expectations for greener processes and products. To meet the green aspirations of our customers, JTC, as Singapore's lead agency for industrial development with over 80% of industrial land under our belt, is well-positioned to take the lead in promoting sustainability in the manufacturing and built-environment sectors.

We proactively galvanise businesses to adopt environmentally sustainable solutions within our estates. Efforts in our existing estates are directed towards generating renewable energy, enhancing greenery and biodiversity, encouraging waste reduction and circular economies and achieving energy and water efficiency. We also ensure that our construction processes are sustainable when developing new estates.

2.   What role does digitalisation play for JTC today?

Digitalisation plays a big role in the planning, design, construction and management of our industrial spaces.

For example, Building Information Modelling (BIM) is used in the master planning and designing of our new flagship estates, Punggol Digital District and Jurong Innovation District. This allows data collected throughout the building project to be analysed for materials trackability, construction productivity and to reduce construction wastage.

In the past two years, we have also incorporated Integrated Digital Delivery (IDD) extensively into the construction phase and have been working with industry partners to trial and develop new digital use cases. IDD integrates digital technologies such as

C. Chung (✉)
Group Director of Engineering, JTC, Singapore, Singapore
e-mail: Calvin_CHUNG@mnd.gov.sg

© Springer Nature Switzerland AG 2022
O. Inderwildi and M. Kraft (eds.), *Intelligent Decarbonisation*, Lecture Notes
in Energy 86, https://doi.org/10.1007/978-3-030-86215-2_16

BIM, virtual reality and photogrammetry (a technique to create 3D-like images) for indoors monitoring as well as drones for aerial site monitoring to enable the project team to perform data-driven decisions.

It is also worth noting that JTC is developing the 50-hectare Punggol Digital District (PDD), a smart and sustainable mixed-use district that houses a creative ecosystem and is a living lab for cybersecurity, smart living and smart estates solutions. Opening in phases from 2023, the PDD will be Singapore's first business park to offer businesses "plug and play" digital infrastructure through the Open Digital Platform (ODP). As a secured platform connected to a network of sensors and systems, ODP will collect data such as building data (e.g. lifts, lighting, mechanical and electrical systems, occupancy) and environmental data (e.g. temperature, rainfall) which will be utilised and integrated to optimise operations and facilities management.

3.   **Do you think digitalisation has significant potential for reducing CO_2 emissions in Singapore? If yes, what are the areas of interest and by how much?**

Digitalisation does indeed have the potential to help us drive sustainability. Besides cutting down on working hours and paper usage, it allows us to collect data that can be analysed and used to develop ways to achieve higher productivity and resource efficiency. Digital tools such as the smart facilities management systems housed at the J-Ops Command Centre allow JTC to centrally and remotely monitor, analyse and optimise the performance of our buildings to reduce energy consumption in our industrial estates. This helps to reduce our active operation carbon.

Digital construction solutions, on the other hand, help to reduce the waste generated during our construction processes and therefore reduce our embodied carbon.

4.   **What R&D would you like to see in academia that can support JTC's efforts?**

JTC welcomes academia to propose R&D related to digitalisation, automation and sustainability as these will be key areas of focus for us in the next few years.

5.   **If there was one problem that you could solve with the use of AI, what would it be?**

AI has advanced by leaps and bounds in recent years, for example in visual and speech recognition, and we are hopeful that AI can be extended to help improve automation and reduce manpower reliance, especially during the construction stage. The challenge in construction is always the complexity of the work and the highly varied site situation and conditions that the robots and machines have to adapt and react to. AI can also help to improve the speed of building planning and design by automatically generating parametric elements for the designers and engineers.

Calvin Chung was JTC's Group Director of Engineering and Director of the Future of Buildings and Infrastructure Division. He oversees engineering design and drives innovation in the building and infrastructure sector to enhance quality, safety and efficiency. Given that sustainability solutions often require an engineering perspective to be brought to life, Calvin in 2019 was also appointed the Chief Environment Officer in JTC as a natural extension of his role as Group Director of Engineering.

Chapter 17
Responsive Carbon Neutral Settlements

Gerhard Schmitt, Heiko Aydt, Jimeno A. Fonseca, Juan Acero, Jan Perhac, and Ido Nevat

Abstract Carbon neutrality could become a necessary condition for human settlements to attain liveability, sustainability and resilience. For most cities, this implies the need for decarbonisation. The replacement of fossil fuels with clean energy not only reduces greenhouse gas emissions, but also improves outdoor and indoor air quality. In urban centres that primarily rely on fossil sources for electricity production in or close to the city, decarbonisation could also reduce the Urban Heat Island (UHI) phenomenon, which poses a severe threat to tropical and subtropical cities. Cities emit more than two-thirds of global greenhouse gases, yet they have the potential to achieve carbon neutrality or even become carbon negative. Towns, villages and low-density settlements can transition from being fossil energy consumers to renewable energy producers, thereby reducing the global carbon footprint of urban regions. To implement such changes, city governments need to develop convincing holistic data-driven scenarios through simulations. For this purpose, we build a digital urban climate twin, using a federation of coupled models. We start with climate, transportation and energy models, followed by socioeconomic models for a holistic simulation. In parallel, we consider technological advances, social responsiveness and fair Artificial Intelligence (AI), which are crucial in achieving carbon-neutral settlements.

17.1 Introduction

Smart city agendas consider the expected positive impact of Artificial Intelligence (AI) and digital technologies on carbon emissions and related costs of buildings,

G. Schmitt (✉) · H. Aydt · J. A. Fonseca · J. Perhac
Singapore-ETH Centre, 1 CREATE Tower, Singapore, 138602, Singapore
e-mail: schmitt@arch.ethz.ch

J. Acero
SMART MIT, 1 CREATE Tower, Singapore, 138602, Singapore

I. Nevat
TUMCREATE, 1 CREATE Tower, Singapore, 138602, Singapore

© Springer Nature Switzerland AG 2022

149

O. Inderwildi and M. Kraft (eds.), *Intelligent Decarbonisation*, Lecture Notes in Energy 86, https://doi.org/10.1007/978-3-030-86215-2_17

neighbourhoods and cities. Examples include embedding AI applications into cooling and heating systems in smart buildings, smart grids and smart grid-enabled heat pumps, which are driving the electrification of thermal processes. Burning fossil fuels in or adjacent to a city is seen less and less as a sustainable option. Increasingly, the production of renewable energy where it will be consumed is becoming a major source of power. At the same time, transnational and transcontinental electricity networks are gaining traction.

Today, more than two-thirds of the world's megacities and most of the rapidly urbanising settlements are found in the tropical and subtropical regions, where fossil fuels are still a primary source of energy. Fossil fuels have enabled and powered the world since the industrial revolution. They helped to lift millions out of poverty and provided amenities in daily life that were unknown before. Back when the global population was a fraction of today's, the release of CO_2 gasses into the atmosphere was not seen as a problem. Yet the exponential growth of emissions has led to a continuous acceleration of global climate change. Cities, covering less than 3% of the Earth's surface, account for more than two-thirds of greenhouse gas emissions, on the pretext that high-density urban environments need concentrated sources of energy, such as fossil fuels, to stay efficient.

The scope of this chapter includes cities and other human settlement types. As of 2020, four out of ten people live in rural settings, and they contribute to climate change as well as do activities including agriculture, long-distance ground transportation, air and sea transportation. Almost all of these sectors depend on fossil fuels. Moving towards electrification and stepwise decarbonization is therefore necessary. Moreover, towns, villages, and new urban–rural settlement types can evolve into food, recreation and renewable energy providers to power future cities.

Every settlement has its own energy supply and demand profile. Therefore, the potential for CO_2 emission reduction differs considerably across regions, countries and climatic zones. The path towards carbon neutrality takes into account the different characteristics of settlement types such as cities, towns or villages and their common goal to improve the quality of life and health of their inhabitants, and sustainability and resilience for future generations. The emergence of a new knowledge-based economy, starting with the energy, manufacturing and transportation sectors, will not only support this change, but also create jobs in sustainable sectors. To illustrate our hypothesis, we compare Switzerland and Singapore, two countries that, although in different climate zones, have much in common.

17.2 A Holistic View on Settlements

Cities are complex cyber-physical and social structures. They are a concentration of people, materials, knowledge, finances, crime, inequalities and opportunities, of which some can be quantified better than others. Cities follow universal scaling laws (West 2017), which make part of their development, function and behaviour predictable. Governance, cultures, religions, expectations and attractiveness are as important, but are more difficult to quantify.

Moving from this abstract level to specific places exposes major dissimilarities between settlements in different climate zones, development states, economic conditions and cultures. Yet, people in most of them share a common desire for high liveability (City in Your Hands, Gerhard Schmitt, E. Tapias, and M. H. Wisniewska, eds. 2019). Liveability is the combination of job opportunities, health, safety, security, and good education. Good indoor and outdoor air quality, outdoor thermal comfort, less noise and green spaces are as important for high liveability.

The Maslow pyramid describing a hierarchy of human needs was often used in the past to differentiate basic physiological needs and high-level aspirations such as self-actualisation. Yet we see that there is no clear separation between these levels, and that the requirements are in a networked, rather than a hierarchical order with each other. We therefore model settlements as complex systems, made for the benefit of the humans that live in them.

In a complex settlement system, the human should not be in the centre of observation, like in the first smart cities–but in the centre of action, like in direct democracies. This has influence on the way we model and simulate the behaviour of such settlements. In the past, hierarchical and low dimensional models were used in the simulation and planning of cities. We propose a federation of models that better reflects the processes occurring in settlements, and therefore can help to better simulate scenarios of possible liveable, healthy and resilient settlements.

In this federation of models, energy, and in particular fossil energy, has a pivotal role, because it has so many interactions with other processes in the settlement: with employment, industry, transportation, housing, heat and health. And as decarbonization would be a major change in energy supply, it will influence all those related areas. Only if we consider the effects and side effects of decarbonization in a federation of models, can we propose realistic decarbonisation scenarios that are beneficial.

17.3 Responsive Cities and Carbon Neutrality

Responsive cities evolve from smart cities, with a fundamental difference: The citizens move from the centre of attention to the centre of action. Responsive citizens use smart technology to contribute to planning, design and management of their cities (City in Your Hands, Gerhard Schmitt, E. Tapias, and M. H. Wisniewska, eds. 2019; EUROPEAN CITIZENS 2020). Responsive cities bring the city back to their citizens. With this, the citizens' own responsibility becomes the foundation of the responsive city. Cities evolve from being smart to being responsive. As citizens use smart city technology, they will be able to closely follow the functioning of a city. Transparency, visualizing data and the positive consequences of fossil energy free settlements will be necessary for urban governments to convince citizens to support the decarbonisation measures. A basis is the emerging concept of Citizen Design

Science, a combination of Citizen Design, Citizen Science and Design Science is under development (Mueller et al. 2018).

Carbon Neutrality is a strategic societal goal to help mitigate climate change. It consists of a state where individuals, companies, and institutions balance their carbon footprint with actions to reduce their emissions (carbon reduction) while contributing to an effective reduction elsewhere (carbon offsetting) (Coulter et al. 2008). Today, more and more cities and nations pledge to reach Carbon Neutrality before 2050 (Countries | Climate Action Tracker 2020) such as the recent examples of New York City (McKinley and Plumer 2020) and the UK (Skidmore 2020).

The first step towards a carbon-neutral Singapore is the accountancy of carbon emissions. Today, carbon emissions in Singapore are estimated based on default emission factors provided in the revised 1996 IPCC Guidelines. This approach estimates the carbon emissions from a production perspective, accounting only for the emissions produced within the country's territory. Another plausible approach is to evaluate carbon emissions from a consumption perspective. In this approach, the emissions are attributed to the goods and services purchased by local consumers (Pichler et al. 2017). As global production chains become more extensive and sophisticated, the difference between these two approaches becomes significant. In the UK, five different studies concluded that emissions due to consumption are up to 40% higher than due to production within the country (Failed 2008). Should this situation also apply to Singapore, the effectiveness of climate mitigation policies based on territorial emissions alone may not be sufficient (Schulz 2010).

A second step towards a Carbon Neutral Singapore will entail the assessment of potential pathways for decarbonization. Here a cross-sectoral view of the benefits, trade-offs, and synergies of decarbonization strategies will be necessary. Due to limited access to land and renewable energy resources, Singapore, as well as many other industrialized nations, has placed a strong bet on energy efficiency to tackle climate change (Secretariat 2016). Technological advances in the industry, transportation, and building sectors are indeed important measures to mitigate carbon emissions. However, new opportunities from the point of view of consumer behaviour need to be explored to reach carbon neutrality. Examples include alternative alimentary products, fair-trade clothing, non-plastic packaging, building material circularity, and changes in traveling and energy consumption behaviour. The challenge lies in how to balance behavioural changes with technological advances for the benefit of a prosperous and yet carbon-neutral economy.

17.4 Anthropogenic Heat Fluxes in Singapore

According to Fig. 17.1 Singapore produces close to 20 Mtoe of anthropogenic heat emissions every year (Kayanan et al. 2019). This is equivalent to the combustion of close to 140 million barrels of oil every year and similar to the entire oil reserves of countries like Germany (145), Poland (138), and Cuba (124). The anthropogenic heat emissions of Singapore are generated by five key contributors. These are industry

(58.5%), power generation (15.3%), buildings (11.9%), transportation (11.4%), and human metabolism (2.8%).

Most sources of anthropogenic heat in Singapore emit heat into the environment (air and sea) at different intensities throughout the City. These intensities or heat fluxes can depend on the location of the source. Figure 17.2 describes this phenomenon for buildings (Resende Santos et al. 2020). As it can be seen, anthropogenic heat holds a close connection to the location of the emitting source. In the case of buildings, heat fluxes also vary for subgroups of buildings. For example, commercial areas, due to their intense economic activity and energy consumption, generate several times more heat than residential areas. As a result. A key hotspot of heat flux for the building sector is, among others, the business district of the city.

On the other hand, heat fluxes in Singapore do not only vary based on location. Since heat fluxes are highly connected to the daily operations of the source, they also vary throughout every hour of the day. Figure 17.3 describes this phenomenon for the transportation sector of Singapore (Ivanchev and Fonseca 2020). As it can be seen, anthropogenic heat fluxes are directly connected to the activities of this sector. In particular, hotspots of heat fluxes in the transportation sector of Singapore will be present in both areas of congestion as well as during peak hours for traffic. As a result, it can be concluded that Anthropogenic heat fluxes in Singapore hold a tight connection to where and when activities in diverse economic sectors take place.

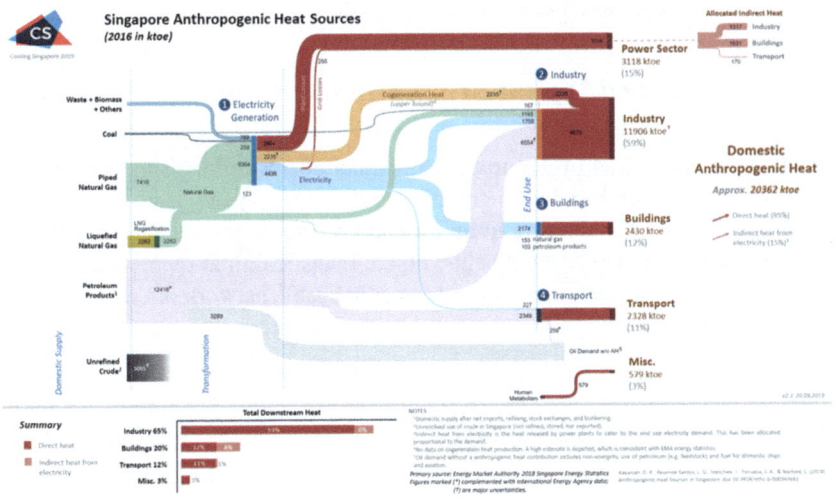

Fig. 17.1 Sources of anthropogenic heat in Singapore. Copied with permission of the authors (Kayanan et al., 2019)

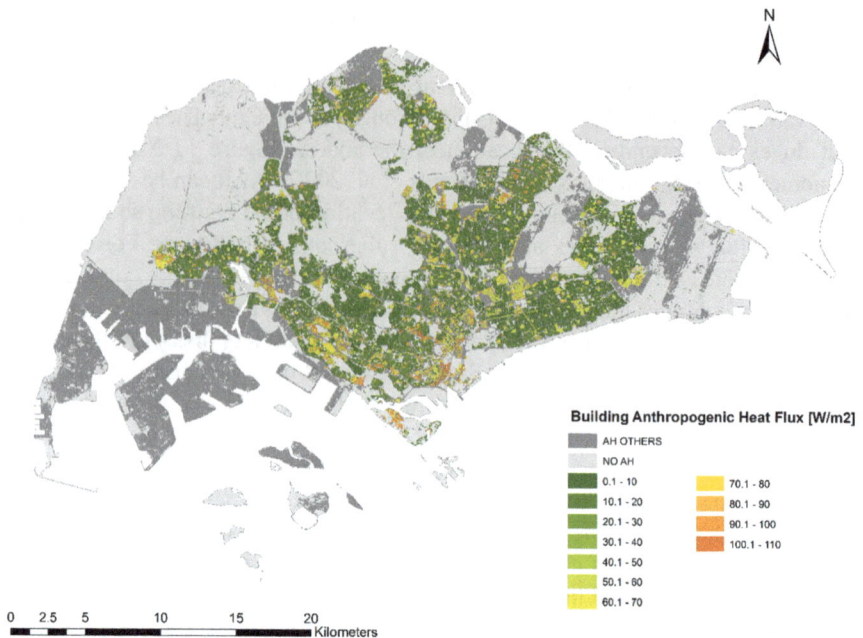

Fig. 17.2 Spatial distribution of anthropogenic heat flux due to buildings in Singapore. Copied with permission of the authors (Resende Santos et al. 2020)

17.5 The Planned Digital Urban Climate Twin

For a holistic simulation of the urban heat island effects no one single model will suffice. In addition, embedding the reduction of anthropogenic heat in the city into an overall strategy of increasing liveability and resilience through decarbonization showed the need for two developments: first, a realistic digital representation of all relevant features of a settlement; second, an expandable network of coupled models to perform simulations of desirable scenarios for the future city.

17.5.1 A Federation of Models

The Digital Urban Climate Twin (DUCT) is a subset of a digital urban twin. It is a "set of specialised urban climate models and anthropogenic heat emission models needed to closely resemble the urban climate dynamics for a particular city of interest. It allows its users to experiment with a digital representation of the city, its urban climate and the various anthropogenic contributors to urban heat (e.g., industry, traffic, and buildings)" (Failed forthcoming). From a technical point of view, a DUCT can be realised as a federation of coupled models that automate the workflows of

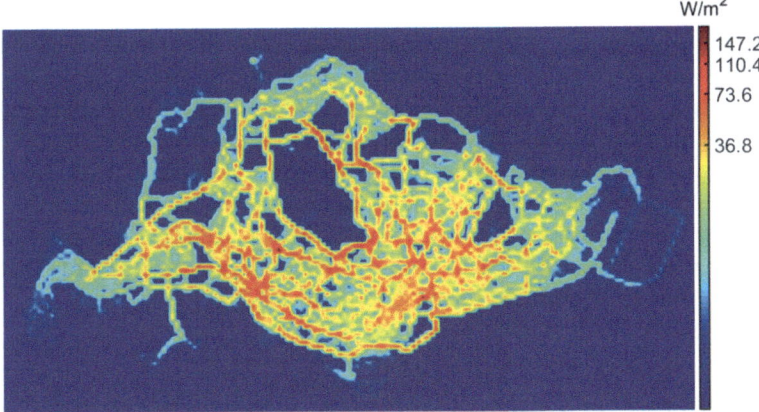

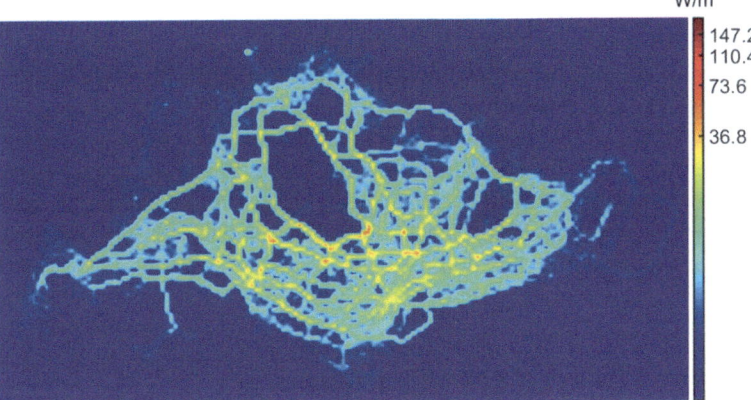

Fig. 17.3 Spatio-temporal distribution of anthropogenic heat flux due to transportation in Singapore. Top: 7–8 am. Bottom: Noon. Copied with permission of the authors (Ivanchev et al. 2020)

individual steps (e.g., data pre-processing, simulation, data analysis) needed in order to evaluate scenarios for urban heat mitigation (Failed 2020). Each scenario may address a different 'what-if question', such as "What if electromobility is used? What if district cooling is used?".

The DUCT is an essential tool for urban climate design and management. "Urban climate design and management refers to the ability to understand the climate science, to modify and maintain the urban climate (temperature, humidity and air-flow) on different urban scales (e.g., island-wide and building-scale), and to comprehend the social science of risks and mitigation to set targets and desired conditions accordingly" (Failed 2020). The DUCT can determine the combination of mitigation measures that is best suited to achieve desirable outcomes for the urban climate. This is done by means of a design-simulation loop that would allow its users to define, simulate and evaluate what-if scenarios—each representing a different combination of heat mitigation measures and assumptions—in an iterative process until a set of promising what-if scenarios has been identified. Further analysis of these scenarios could result in a more informed decision-making process before the implementation of any heat mitigation measures.

The Digital Urban Climate Twin (DUCT) will be developed in Cooling Singapore 2.0, a 3-year project, funded by the National Research Foundation of Singapore. This DUCT would give Singapore's planners and decision makers an important tool that supports climate-informed decision making, representing a step towards enabling Singapore to manage its urban climate through design.

17.6 The Role of AI to Achieve Carbon Neutrality in Cities

Artificial Intelligence (AI) has played a role in architecture and urban design as early as in the 1970s. Due to the lack of suitable algorithms and computational power, expectations could not be met and research moved towards the possible use of AI in the design process instead. With the emergence of smart and responsive cities in the first decade of the new millennium and the dramatic growth in computing power, the role of AI gained traction in design. Research results started to percolate into practice, and in combination with big data, researchers developed a wealth of applications (Koenig et al. 2020). In the federation of models, AI methods will have a growing impact and play a central role.

17.6.1 The Role of AI in Cooling Singapore

Developing a holistic approach to settlement decarbonization towards carbon neutrality requires a multidisciplinary approach due to the complexity of the problem. As such, it requires the development and integration of multiple models that communicate, share and propagate information in a meaningful way–this is what the DUCT does. Achieving this goal requires significant computational power, which is impractical in cases where real-time analysis is required, especially when many what-if scenarios (projections) need to be evaluated.

To alleviate this problem, Machine Learning (ML) methods and algorithms have been developed and implemented during the last few decades by different research communities. ML-based algorithms can be used to solve many problems faced in climate modelling. These include: (1) parametrization of the climate models using advanced regression and classification models (Bechtel and Daneke 2012); (2) providing efficient solutions of complex mathematical equations that result from the first principle physics using Deep Learning methods (Raissi and Karniadakis 2018); (3) replacing complicated physics models with a simple, yet accurate, approximation of those equations with surrogate models (Edeling and e. al, 2018); (4) data fusion of information from multiple sources of information with varying degrees of accuracy, for example, fusion of high-quality data from sparsely deployed weather stations, with dense low-quality crowdsourced data (Zhang and e. al., 2018; Xiang et al. 2020).

One of the challenges that projects such as Cooling Singapore face is the knowledge transfer of advanced ML techniques from theory to practice. ML techniques include Random Forest Regression and Classification methods, deep learning and Gaussian Processes models and Reinforcement Learning.

In spite of best efforts, each of the models contains estimation errors that which stem from multiple approximations, partial and distorted data-sets to list a few. It is important to acknowledge the cascading effect of error propagation throughout the system. This effect may lead to an overall accumulated error and render the system (DUCT) useless.

To address the error propagation challenge, a systematic analysis will be carried out in order to incorporate and fuse those uncertainties into a decision-making framework to help decision makers choose the optimal policy. This involves the development of a principled approach to decision making under uncertainty (which we take in Cooling Singapore). This approach includes the celebrated Expected Utility Theory (Morgenstern and Neumann 1953) as well as other well-established theories, such as Modern Portfolio Theory (Markowitz 1991). In the context of climate-informed urban design, the Cooling Singapore team has developed such a statistical decision-theoretic framework for urban design that incorporates important design criteria such as Outdoor Thermal Comfort (OTC) as well as social and economic indices (Nevat et al. 2020a, 2020b; Zhong et al. 2019).

17.7 The Role of Visualization

Visualization is a key component in design and science. It helps scientists and designers to receive visual feedback on the result of their actions, in addition to numbers and formulas (Burkhard 2008). Visualization can also be a crucial help in communicating scientific results to laypeople and to the public. Visualization can create a common ground for citizens, scientists and governments in the process of decarbonising settlements and reaching carbon neutrality.

17.7.1 Dedicated Visualizations

Cooling Singapore has used advanced computer graphics techniques to communicate its findings with target audiences via dedicated visualizations. Some of these techniques consisted of overlapping spatial and temporal data layers within a simplified 3D environment using game engines (Cristie et al. 2015). Other techniques consisted of using the engines for visualizing results of computational fluid dynamics models within complex 3D environments. Cooling Singapore also tapped into Virtual Reality (VR) to allow users to visually experience thermal comfort levels produced by different types of buses in neighbourhoods of Singapore (Fig. 17.4).

To help audiences to better grasp complex spatio-temporal phenomena such as UHI (Fig. 17.5) and active anthropogenic heat from transportation (Fig. 17.6) on a city scale level we use an interactive, visual storytelling platform with enables direct, low-level access to Graphical Processing Units (GPU) for transformation of digitized spatial and temporal data (Zhong et al. 2019). The low-level access allowed smooth visualization of hundreds of thousands to millions of temperature records without the need to pre-process the data.

The focus then shifted to empowering users to holistically study and improve Outdoor Thermal Comfort (OTC) via an interactive, user-oriented Decision Support System (DSS) (Nevat et al. 2020b), as shown in Fig. 17.7. The DSS allows users to set different input parameters to observe climatic impact, scores and spatial temporal performances of selected individual scenarios for OTC analysis. Parameters include scenarios (e.g. elevated podiums, void decks), weather conditions and climate variables, and indexes such as Physiological Equivalent Temperature (PET) and Mean Radiant Temperature (MRT), among others. This analysis allows users to assess cost benefits, energy consumption and the reduction of CO2 emissions of selected scenarios. With up to over 8,000 possible combinations of parameters, emphasis during the DSS development was also on using User Experience (UX) techniques like the Double Diamond method to create an intuitive yet powerful user interface (UI). The next stage in visualization will be a city-scale DUCT visualization to allow users to study and experiment with a digital representation of the entire city.

Fig. 17.4 Virtual reality bus stop—Simulation and visualization of heat rejection by buses (Michael Joos 2017)

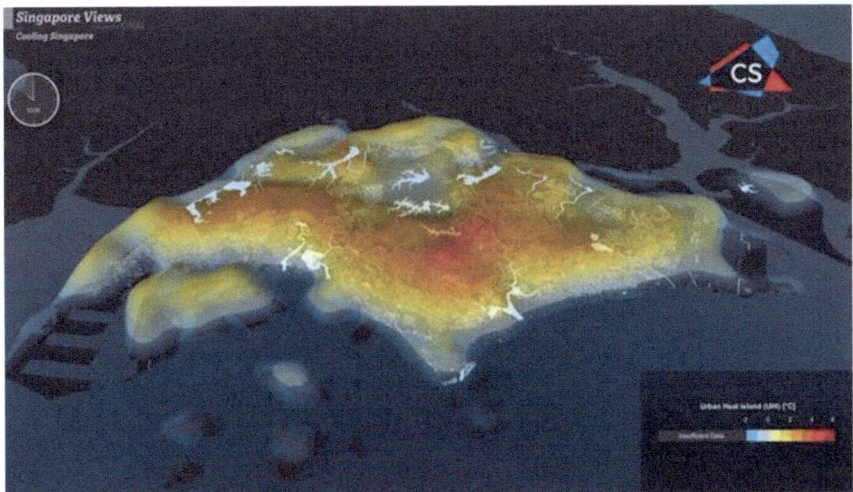

Fig. 17.5   Urban Heat Island Intensity in Singapore at midnight (Zhong et al. 2019)

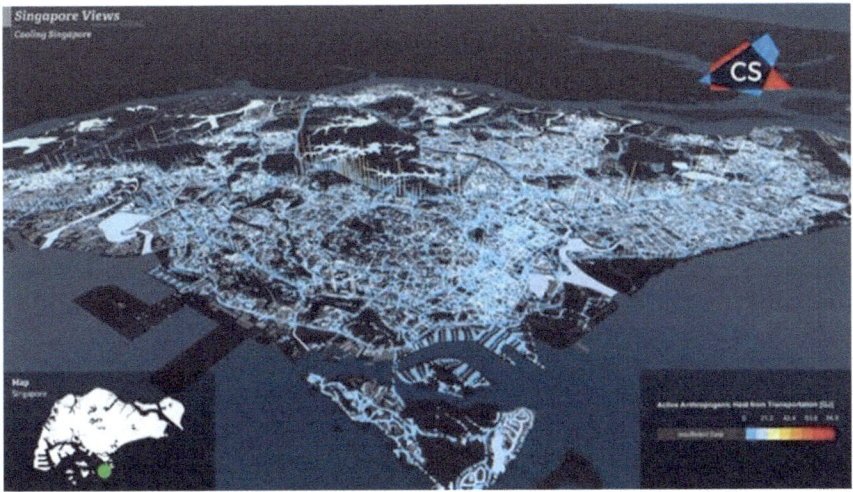

Fig. 17.6   Active Anthropogenic Heat From transportation (Zhong et al. 2019)

17.8   Conclusions

Achieving carbon neutrality for cities, towns and villages around the world is a way to reduce greenhouse and toxic gas emissions, and to improve air quality and outdoor thermal comfort. Using a Digital Urban Climate Twin, we will be able to simulate and assess different scenarios for each settlement, based on its specific profile in terms of climate, altitude, citizen behavior and cultural preferences. Climate change

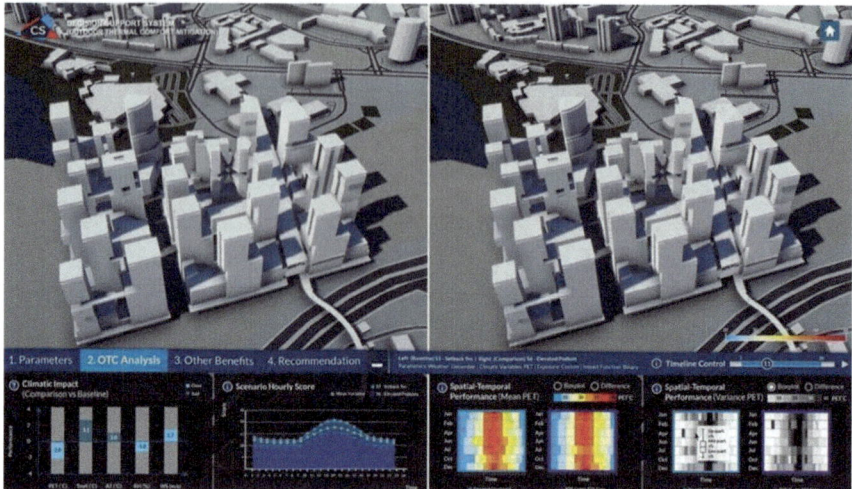

Fig. 17.7 Decision Support System comparing two scenarios for OTC assessment developed during CS1.5 (Cong Ye, Michelle Chan 2020)

in the next decades will lead to less heating demand and CO_2 emissions in parts of the world, while the number of cooling degree days will increase and thus the cooling demand will increase everywhere. Globally, there could be a net CO_2 reduction over one year. But in Singapore and other tropical cities where no heating is needed, there will be no beneficial balancing effect of global warming, and strong measures have to be taken to reverse further increase of the urban heat island intensity.

The more cities depend on fossil fuel energy today, the higher will be the positive effects of decarbonisation towards carbon neutrality, such as better air quality and a lower UHI effect. It might in fact lead to cities and settlements cooling down faster, although the result of less need for cooling, higher efficiency in industrial processes and renewable energy production on building roofs and façades are not known and will be the subject of future research. Decarbonisation towards carbon neutrality will make settlements more liveable and resilient and life more comfortable and healthier in cities in all climates, due to the reduction of emissions.

17.9   Outlook

To achieve the goal of carbon neutrality of human settlements, the support of both the population and the government, aided by improvements in hardware and software, is necessary. Hardware improvements include optimization of and new building materials with radically less CO_2 embodiment and high albedo values, integrated renewable energy and food production in close proximity to the location of consumption, more efficient and intelligent industrial production and products. Software improvements are connected to these developments and include improved data analytics, close to real-time predictive simulations, and fair AI embodiment in all design and simulation programs. Decarbonization towards carbon neutrality will only work if the population supports this future with a responsive attitude, meaning that it can participate in data collection, design proposals and city management. Combining these possible futures, and compared to today, a carbon–neutral Singapore could reduce its emissions by more than 50% in 2040 with a long-term yearly net reduction of expenses by several billion dollars.

Acknowledgements This research is supported by the National Research Foundation, Prime Minister's Office, Singapore under its Campus for Research Excellence and Technological Enterprise (CREATE) programme.

References

Aydt H (2020) Cooling Singapore–towards urban climate design and management. In: Cairns S, Tunas D (eds) Indicia 03

Aydt H (2020) Towards a digital urban climate twin: Simulation-as-a-Service (SaaS) for model integration, in "ETH Research Collection," Singapore-ETH Centre, Singapore

Bechtel B, Daneke C (2012) Classification of local climate zones based on multiple earth observation data (2012). IEEE J Sel Top Appl Earth Obs Remote Sens 5(4):1191–1202

Burkhard R (2008) Knowledge visualization: the use of complementary visual representations for the transfer of knowledge. A model, a framework, and four new approache, doctoral thesis

City in Your Hands (2019) In: Schmitt G, Tapias E, Wisniewska MH (eds) Swiss Federal Institute of Technology in Zurich (ETHZ), Department of Architecture, Chair of Information Architecture, p 258

Cristie V, Berger M, Bus P, Kumar A, Klein B (2019) CityHeat: visualizing cellular automata-based traffic heat in Unity3D. SIGGRAPH Asia 2015 visualization in high performance computing, pp 6–10. https://doi.org/10.1145/2818517.2818527

Countries | Climate Action Tracker (2020). https://climateactiontracker.org/countries/ (accessed Oct 27)

Coulter L, Canadell JG, Dhakal S (2008) Global carbon project: carbon reductions and offsets, in Earth System Science Partnership Report No. 5. Global Carbon Project Report No. 6, Canberra

EUROPEAN CITIZENS INITIATIVE (2020). European Commission. https://europa.eu/citizens-initiative/_en (accessed Oct 30)

Edeling WN et. al (2018) Bayesian predictions of Reynolds-averaged Navier–Stokes uncertainties using maximum a posteriori estimates. AIAA Journal 56(5):2018–2029

Ivanchev J, Fonseca JA (2020) Anthropogenic heat due to road transport: a mesoscopic assessment and mitigation potential of electric vehicles and autonomous vehicles in Singapore, in "ETH Research Collection," Singapore-ETH Centre, Singapore. [Online]. Available: https://doi.org/10.3929/ethz-b-000401288

Kayanan D, Santos LR, Ivanchev J, Fonseca J, Norford LK (2019) Anthropogenic heat sources in Singapore, in "ETH Research Collection," Singapore-ETH Centre, Singapore. [Online]. Available: https://doi.org/10.3929/ethz-b-000363683

Koenig R, Miao Y, Aichinger A, Knecht K, Konieva K (2020) Integrating urban analysis, generative design, and evolutionary optimization for solving urban design problems. Environ Plan B: Urban Anal City Sci 47(6):997–1013. https://doi.org/10.1177/2399808319894986

Markowitz HM (1991) Foundations of portfolio theory, *of portfolio theory*. J Financ 46(2):469–477

McKinley J, Plumer B (2020) New York to approve one of the world's most ambitious climate plans. The New York Times. https://www.nytimes.com/2019/06/18/nyregion/greenhouse-gases-ny.html (accessed Oct 27)

Morgenstern O, Neumann JV (1953) Theory of games and economic behavior. Princeton university press

Mueller J, Lu H, Chirkin A, Klein B, Schmitt G (2018) Citizen design science: a strategy for crowd-creative urban design. Cities 72(A):181–188. https://doi.org/10.1016/j.cities.2017.08.018

Nevat I, Ruefenacht LA, Aydt H (2020) Recommendation system for climate informed urban design under model uncertainty. Urban Clim 31. https://doi.org/10.1016/j.uclim.2019.100524

NevatI I, Pignatta G, Rufenacht L, Acero JA (2020b) A decision support tool for climate-informed and socioeconomic urban design. Environ Dev Sustain. https://doi.org/10.1007/s10668-020-009 37-1

Pichler P, Zwickel T, Chavez A, Kretschmer T, Seddon J, Weisz H (2017) Reducing urban greenhouse gas footprints, in "Scientific Reports, 7(1):14659. [Online]. Available: https://doi.org/10.1038/s41598-017-15303-x

Raissi M, Karniadakis GE (2018) Hidden physics models: machine learning of nonlinear partial differential equations. J Comput Phys 357:125–141

Resende Santos LG, Singh VK, Mughal MO, Riegelbauer E, Fonseca JA, Norford L, Nevat I (2020) Copyright – Non-Commercial Use Permitted. Cooling Singapore (CS). https://doi.org/10.3929/ethz-b-000440490

Schulz NB (2010) Delving into the carbon footprints of Singapore-comparing direct and indirect greenhouse gas emissions of a small and open economic system. Energy Policy 38(9):4848–4855. [Online]. Available: https://doi.org/10.1016/j.enpol.2009.08.066

Secretariat N. C. C (2016) Singapore's climate action plan: take action today, for a carbon-efficient Singapore. Accessed: October 29, 2020. [Online]. Available: https://www.nccs.gov.sg/media/pub lications/climate-plan-take-action-today-for-a-sustainable-future

Skidmore C (2020) UK becomes first major economy to pass net zero emissions law. https://www.gov.uk/government/news/uk-becomes-first-major-economy-to-pass-net-zero-emissions-law (accessed Oct 27)

West GB (2017) Scale: the universal laws of growth, innovation, sustainability, and the pace of life in organisms, cities, economies, and companies. Penguin Press

Wiedmann T, Wood R, Lenzen M, Minx J, Guan D, Barrett J (2008) Development of an embedded carbon emissions indicator: producing a time series of input-output tables and embedded carbon dioxide emissions for the UK by using a MRIO Data Optimisation System. Research report to the Department for Environment, Food and Rural Affairs by Stockholm Environment Institute at the University of York and Centre for Integrated Sustainability Analysis at the University of Sydney. Defra

Xiang Q, Nevat I, Peters GW (2020) Bayesian spatial field reconstruction with unknown distortions in sensor networks. IEEE Transactions on Signal Processing 68:4336–4351

Zhang P et. al (2018) Spatial field reconstruction and sensor selection in heterogeneous sensor networks with stochastic energy harvesting. EEE Trans Signal Process 66(9):2245–2257

Zhong S, Nevat I, Acero JA, Rüfenacht LA, Jan P, Koh E (2019) A novel decision support tool for climate-responsive urban design. In: Journal of Physics: Conference Series, CISBAT 2019, Climate Resilient Cities—Energy Efficiency & Renewables in the Digital Era, vol 1343, 4–6 September 2019, EPFL Lausanne, Switzerland Published under licence by IOP Publishing Ltd

Zhong S, Nevat I, Acero JA, Rüfenacht LA, Perhac J, Koh E (2019) A novel decision support tool for climate-responsive urban design. CISBAT. https://doi.org/10.1088/1742-6596/1343/1/012011

Chapter 18
Decarbonising Transport with Intelligent Mobility

Justin D. K. Bishop

Abstract Intelligent mobility is the smarter, greener and more efficient movement of people and goods. Road transport accounted for 24% of global greenhouse gas emissions in 2019. Decarbonising the transport system requires decision-makers to adopt a 'vision and validate' approach to promote active and sustainable travel modes and move away from schemes that lock in private vehicle use. Creating a digital twin requires models representing demand for transport, supply of transport options, the network which links origins to destinations and the decisions made by consumers. Mode choice and trip purpose can be inferred using artificial intelligence techniques which incorporate data from the full range of conventional and novel, static and continuous data sources. Digital twins developed on the basis of transparency and inclusion present a unique opportunity to engage with the public in different ways to deliver better outcomes.

18.1 Introduction

Intelligent mobility is focused on enabling the smarter, greener and more efficient movement of people and goods (Mulley 2019). Mobility, or the ability to move easily between two or more points at a specific time for a purpose, is satisfied through a transportation network which supports: active modes, such as walking and cycling; sustainable modes, such as public transport (buses and trains); and private transport, such as cars and taxis. Road transport accounted for 24% of global greenhouse gas (GHG) emissions in 2019 (IEA 2019a, b). This chapter focuses on the use of data, technology and models to meet the green objective of the intelligent transport system: to support the efforts of individual states to achieve net zero GHG emissions by 2050 (UNFCCC 2019), in line with the Paris Agreement (UNFCCC 2016).

The cyber-physical system (CPS) is an orchestration of various digital technologies to create a digital representation, or 'twin,' of the physical world (Inderwildi et al. 2020). Digital twins of physical assets, such as the transport system, are the means to achieve better-informed decisions and sustainable solutions (Bolton 2018; Wan

J. D. K. Bishop (✉)
Arup, 13 Fitzroy St, London W1T 4BJ, UK

© Springer Nature Switzerland AG 2022 163
O. Inderwildi and M. Kraft (eds.), *Intelligent Decarbonisation*, Lecture Notes in Energy 86, https://doi.org/10.1007/978-3-030-86215-2_18

et al. 2019). Traditionally, transport planning has used data and models to 'predict and provide' the infrastructure needed to meet forecast demand. This has entrenched a car-dominated environment and failed to deliver the best outcomes.

A 'vision and validate' (Jones 2016) approach starts with a vision of what the development seeks to achieve, recognising 'planning for people will result in places for people; planning for cars will result in places dominated by cars' (CIHT 2019). Similarly, a 'decide and provide' methodology highlights the three main characteristics of a transport system: access; resilience; and scale. Access is key for a thriving society through connecting people, goods, services and opportunities. Resilience recognises access to the transport system needs to adapt to meet our behaviour over time. Scale refers to supporting the demand that society believes is appropriate and feasible, rather than relying on predictions (Lyons et al. 2014).

An important first step in creating a digital twin is to appreciate the spatial separation between homes, businesses, schools, industry, leisure activities and other places where travel might begin or end. The size of this spatial mismatch, or separation, can constrain our mobility options. In Europe (Rojas-Rueda et al. 2016) and the USA (ORNL 2017), average walking and cycling commuter distances are up to 1.4 km and 4.4 km, respectively. However, the scale of additional walking and cycling trips is large since 46–50% of car trips in Europe (WHO 2020) and the USA (ORNL 2017) cover a distance of less than 5 km. Over 5 km, motorised public transport modes may be required to maintain acceptable journey time and comfort. The greater the spatial mismatch, the higher the proportion of trips that must be satisfied using motorised modes, locking in a minimum energy use.

Minimum energy use constrains the decarbonisation potential of the transport system. More efficient traffic management can only reduce emissions by a fraction of what could be achieved if those vehicles were not circulating in the first instance. Switching from fossil fuels to low emissions alternatives, including electricity, shifts emissions further up the fuel supply chain away from the vehicle tailpipe. However, many alternative fuels depend on more energy-intensive production pathways than those used to deliver conventional petrol and diesel. The result is some alternative fuels fail to deliver lower well-to-wheel GHG emissions than traditional road fuels. The energy intensity of producing alternative fuels, with associated impacts on space, water and other natural resources, raises broader sustainability questions (Bishop et al. 2012). Importantly, switching to low emissions fuels does not address the other challenges which car-dominated cities pose to the environment and society, such as air pollution, congestion, noise and road safety.

18.2   Creating a Digital Twin

Creating a digital twin requires models representing both the demand for transport and the supply of transport options. Demand is represented by the spatial distribution of the population, while the supply reflects the various modal networks linking origins and destinations.

Mode choice is influenced by enabling infrastructure. Active transport infrastructure must be designed with pedestrians and cyclists in mind, in the same way as roadways are designed to enable the safe and efficient movement of vehicles. Increased physical activity associated with uptake of active transport modes has a positive impact on public health. This is an important co-benefit, as average health spending in OECD countries accounted for almost 9% of gross domestic product and 15% of total government spending in 2018 (OECD 2020).

Modelling and simulating decisions made by consumers may be the most difficult aspect of building a digital twin. Trip purpose and the properties of each transport mode—such as comfort, convenience and cost—influence both the decision to travel and when the trip occurs. It is essential for the digital twin to incorporate the full range of trip purposes because simplified models focusing only on commuters do not account for a large proportion of all travel. For example, commuting and business journeys comprised only 18% of car trips and 29% of person miles travelled in England in 2018 (Department for Transport (DfT) 2019). In the USA, 51% of private vehicle trips and 55% of person miles travelled were associated with commuting between home and work in 2017 (ORNL 2017).

Similarly, the digital twin must recognise explicitly a consumer may have access to different transport options and may not use the same one for all trips. For example, 50% of car trips in England in 2018 were used for leisure and shopping, while 69% of cycle trips were for leisure and commuting (Department for Transport (DfT) 2019).

18.2.1 Data Sources

Traditionally, snapshot surveys have been used to map the demand for transport. For example, the census reflects the state and spatial distribution of a population in various ways, such as by age, gender, education level and household income. Independent data sources, such as the voter database or income tax returns, can provide more up-to-date insights into the changes in the distribution of populations through time (Barbosa et al. 2018). National household travel surveys yield population-level trends, while roadside interviews are smaller in scale, but give more detailed information about the specific trip being undertaken.

Continuous data collection methods have the key advantage of avoiding the need to infer activity outside of the survey period. However, these outputs may still provide imperfect information on trip start, end, purpose, mode and time of day. For example, automatic traffic counters (ATC) return vehicle number, speed, size and direction of travel by road link. Automatic number plate recognition (ANPR) can identify a vehicle's make and model and relate it to a registered owner and address. These data sources offer details of how the transport network is being used by vehicle type and time of day, but cannot provide insight into trip purpose or destination.

Mobile network data (MND) and public transport smart card data have been described as large-scale, opportunistic human mobility sensors (Anda et al. 2016). MND (Barbosa et al. 2018) and satellite navigation routes, allow the trip start and end to be inferred. However, the accuracy of these inferences may be reduced in rural areas with sparsely distributed mobile phone towers. Additionally, satellite navigation data may return incomplete trip information due to urban canyons or when the navigation start and end points do not align with the trip origin and destination. Smart cards used on public transport can extend basic time of day ridership data to include more detailed trip information, including trip start, end and frequency. The fusion of smart card data and MND has been used to derive complementary information on the same trip, particularly the factors influencing route choice (Poonawala et al. 2016).

Movement of goods is an important part of a modern economy, but often overlooked in transport models. Here, the decision to make a trip at a given time is governed by supply chain needs, rather than driver preferences. Network-level information on the numbers of goods vehicles circulating in space and time can be derived from ATCs. However, radio frequency identification (RFID) and GPS tracking of shipments can provide more useful data on the movements of goods, including origins, destinations and transshipment points (de Jong et al. 2016). The Blockchain distributed ledger technology may allow goods to be traced through the supply chain quickly to support models of goods flows (TSC 2018). Integrating inductive loop signatures with weigh-in-motion sensors can be used to infer the vehicle configuration and type of goods carried (Tok and Ritchie 2014).

Access to road infrastructure is relatively unregulated, with low barriers to entry to the general population. In contrast, trains, aircraft and ships operate within a controlled system of dedicated termini, connecting infrastructure and timetables. Additionally, operators are required to maintain accurate records of what and who is being transported. These requirements can reduce, but not eliminate, the data challenge associated with trip origins, destinations, mode and purpose because the public transport leg represents only part of the overall trip in most cases.

18.2.2   The Role of Artificial Intelligence

Artificial intelligence (AI) is well suited to solving transport problems because it can infer relationships across a range of data sources which might be too difficult to achieve using traditional techniques.

Often, AI applications in transport focus on autonomous, connected private or shared vehicles. As with low emissions vehicles, a complete switch from a conventional to an AI-supported fleet will not address the social, environmental and economic externalities associated with car-dominated transport systems.

The widespread adoption of automation may facilitate behaviour change which impacts significantly the amount of energy used to deliver mobility. For example, transport emissions could be 45% lower in a 'have our cake and eat it too' scenario as the full benefits of automation are realised, with few downsides. Under different conditions, emissions could increase by 90% in a 'dystopian nightmare' where motorists embrace full automation and disengage from the responsibilities of driving to undertake other activities. This causes their perceived costs due to traffic to fall, leading to more vehicles in circulation and greater congestion (Wadud et al. 2016). Mitigating the risks associated with unintended consequences requires strong public policy and governance to steer urban digitalisation towards delivering benefits both to society and the environment (Creutzig et al. 2019).

The number and proportion of local, short car trips indicate the scale of activity which could be shifted to active and sustainable modes. While the potential emissions savings are large, achievements in the real world may be modest in the absence of both enabling infrastructure for walking and cycling and attractive public transport alternatives to using private vehicles. Therefore, AI techniques are most valuable for predicting mode choice because infrastructure needs, vehicle types and efficiency simulations all derive from trip knowledge.

Traditionally, discrete choice models have been used to determine mode choice. A growing body of evidence shows a range of data-driven, machine learning techniques outperform traditional statistical approaches (Hagenauer and Helbich 2017), with deep learning methods used to mitigate overfitting (Nam et al. 2017). Importantly, neural networks can predict mode choice following infrastructure changes, such as new public transport options and restrictions to private car use (Buijs et al. 2020).

AI has been used in planning, public transportation, autonomous vehicles and incident detection applications (Abduljabbar et al. 2019). The ability of AI to detect or infer incidents in the transport network is an important feature. Network interruptions may have impacts across multiple time periods, from traffic accidents in the short term to long term road closures due to construction of large infrastructure projects. Such interruptions may affect route choice, mode choice or overall decision to undertake the trip. Equally, the provision of new infrastructure can influence consumer mode choice positively. For example, the Cambridge guided busway was built with dedicated foot and cycle paths. The result was an increase in cycle trips, both by existing cyclists and those induced to use the targeted infrastructure (Panter et al. 2016).

Beyond infrastructure changes, AI can draw on its data sources to infer the sensitivity to financial charges and rewards. For example, high-level continuous data from ATCs can be used to understand the demand response to access charges or car parking charges. However, pairing with ANPR data allows more detailed evaluation of consumer response, by identifying vehicles that pay to access or park. Mapping the addresses where these vehicles are registered to the public transport network may be used to quantify the underserved demand. Integrating with smart card data or MND might provide insight into how those not using the car are satisfying their demand for transport.

18.3 Digital Twin Examples

The digital twin can be used to test decarbonisation strategies associated with either increasing supply of active and sustainable travel mode options or decreasing the attractiveness of using the private car or both. A survey of over 200 car-free city initiatives identified strategies, such as: reviving the social function of streets; improving housing and residential areas; reducing air pollution; raising revenue by internalising the externalities of car use; and rationalising freight management, which focuses on using goods vehicles that are appropriate to the built environment and road network (Ortegon-Sanchez et al. 2017).

A digital twin was developed for Herrenberg, combining an extensive sensor network with models of the built environment, street network, urban mobility patterns and wind flow. The project was driven by a motivation to be transparent and engage the general public. The mathematical street network model made it possible to evaluate the effect of mobility interventions on emissions. Virtual and augmented reality (VR, AR) was used to visualise the impacts of these mobility interventions to support public engagement and multi-stakeholder decision making (Dembski et al. 2020).

Similarly, Virtual Zurich has been established to visualise multi-dimensional models of the city and facilitate public engagement and interaction. The data feeding the city's digital twin is published according to an 'open by default' principle (Schrotter and Hürzeler 2020). Transparency and an open, collaborative approach are seen as key to developing coherent digital twins (Tao and Qi 2019). Zurich's Open Data (available at https://data.stadt-zuerich.ch/) allows users to develop showcases which integrate multiple sources: for example, Fig. 18.1 illustrates a map view that combines a two-dimensional footprint model, a three-dimensional elevation model and a digital terrain model.

An 'outcome-oriented digital twin' was developed for Cambridge to test medium to long term policy scenarios around the commute to work. Choice models for residence location and mode choice drew on conventional and novel data sources, such as the census and ANPR studies. The two scenarios tested were an increase in teleworking and an increase in demand for electric vehicle charging infrastructure. The impacts of these scenarios were evaluated against aims to reduce congestion, to

Fig. 18.1 Zurich MapView based on Zurich's open data, available online at https://data.stadt-zuerich.ch/showcase/zurich-mapview

improve air quality and to upgrade energy infrastructure to support growth and promote sustainability (Wan et al. 2019).

The Chinese government has committed to using digital technologies to rebuild its cities. One example is Alibaba's City Brain pilot project in Hangzhou. City Brain is cloud-based, aggregating data from a range of data sources, such as government, physical sensors and social media and applying AI to propose effective use of resources. For example, City Brain has been used to coordinate traffic and road signals in real-time to reduce congestion significantly: in 2018, City Brain was launched in Kuala Lumpur to manage traffic signals at 281 intersections (Naughton 2020).

18.4 Conclusions

Digital twins are being introduced in cities to facilitate intelligent mobility—the smarter, greener, more efficient movement of people and goods. Developing a digital twin of the transport network requires fusing conventional and novel sources of transport activity data with transport mode network maps. However, the growth in the volume of data from a range of sources does not give us perfect information about consumer behaviour: trip start, end, purpose, mode choice and the underlying factors influencing travel decisions. AI has a key role in filling these gaps in our knowledge: first, to validate the digital twin models of the real system; and second to estimate the response of consumers to 'vision and validate' decarbonisation policy interventions that promote active and sustainable travel over schemes which lock in private vehicle use. The digital twins of Herrenberg, Zurich and Cambridge range in complexity, scale and objective, but share the common themes of transparency, being outcome-oriented and facilitating and enhancing public engagement.

18.5 Outlook

AI and digital technology help decision-makers to understand how and why trips are taken and to infer consumers' response to changes in the transport options available to them. Improving traffic flows and public transport reliability will return GHG emissions savings and favourable outcomes, but is evolutionary in scope and impact. Digital twins can support a step-change in GHG emissions savings by allowing decision makers to evaluate scenarios that achieve large scale shift to active and sustainable transport modes. Importantly, realising this decarbonisation potential requires transport planners to work with land use planners to reduce the minimum transport energy intensity of the built environment.

References

Abduljabbar R, Dia H, Liyanage S, Bagloee SA (2019) Applications of artificial intelligence in transport: an overview. Sustainability 11(1):189. https://doi.org/10.3390/su11010189. https://www.mdpi.com/2071-1050/11/1/189

Anda C, Fourie P, Erath A (2016) Transport modelling in the age of big data. Technical report

Barbosa H, Barthelemy M, Ghoshal G, James CR, Lenormand M, Louail T, Menezes R, Ramasco JJ, Simini F, Tomasini M (2018) Human mobility: models and applications, vol 734. Elsevier B.V. https://doi.org/10.1016/j.physrep.2018.01.001. https://arxiv.org/pdf/1710.00004.pdf

Bishop J, Axon C, Tran M, Bonilla D, Banister D, McCulloch M (2012) Identifying the fuels and energy conversion technologies necessary to meet European passenger car emissions legislation to 2020. Fuel 99:88–105. https://doi.org/10.1016/j.fuel.2012.04.045

Bolton A (2018) The Gemini principles. https://doi.org/10.17863/CAM.32260

Buijs R, Koch T, Dugundji E (2020) Using neural nets to predict transportation mode choice: an amsterdam case study. Proc Comput Sci 170:115–122. https://doi.org/10.1016/j.procs.2020.03.015. https://pdf.sciencedirectassets.com/280203/1-s2.0-S1877050920X00081/1-s2.0-S1877050920304440/main.pdf?X-Amz-Security-Token=IQoJb3JpZ2luX2VjEA0aCXVzLWVhc3QtMSJIME YCIQCA5j4VpWM7NwSpSYk8yfzOP74BvcTkggpOQ2CnHGTv1gIhAOIwnERfLmsTPgApJ Fw1ox02s0zF2OlDhtD8hP7gebGx

CIHT (2019) Better planning, better transport, better places. Chartered Institute of Highways & Transportation. www.ciht.org.uk

Creutzig F, Franzen M, Moeckel R, Heinrichs D, Nagel K, Nieland S, Weisz H (2019) Leveraging digitalization for sustainability in urban transport. Global Sustain 2(e14):1–6 (2019). https://doi.org/10.1017/sus.2019.11. https://doi.org/10.1017/sus.2019.11

de Jong G, Tavasszy L, Bates J, Grønland SE, Huber S, Kleven O, Lange P, Ottemöller O, Schmorak N (2016) The issues in modelling freight transport at the national level. Case Stud Transp Policy 4(1):13–21. https://doi.org/10.1016/j.cstp.2015.08.002. http://eprints.whiterose.ac.uk/89282/1/No_authors_The_issues_in_modelling_freight_V2.pdf

Dembski F, Wössner U, Letzgus M, Ruddat M, Yamu C (2020) Urban digital twins for smart cities and citizens: the case study of Herrenberg, Germany. Sustainability 12(6):2307. https://doi.org/10.3390/su12062307. https://www.mdpi.com/2071-1050/12/6/2307

Department for Transport (DfT) (2019) Purpose of travel—GOV.UK. https://www.gov.uk/government/statistical-data-sets/nts04-purpose-of-trips

Hagenauer J, Helbich M (2017) A comparative study of machine learning classifiers for modeling travel mode choice. Expert Syst Appl 78:273–282. https://doi.org/10.1016/j.eswa.2017.01.057. https://www.researchgate.net/profile/Julian_Hagenauer/publication/313685309_A_comparative_study_of_machine_learning_classifiers_for_modeling_travel_mode_choice/links/5bc2f37a458515a7a9e73413/A-comparative-study-of-machine-learning-classifiers-for-modeling

IEA (2019a) Tracking transport—analysis—IEA. https://www.iea.org/reports/tracking-transport-2019

IEA (2019b) Global CO$_2$ emissions in 2019—analysis—IEA. https://www.iea.org/articles/global-co2-emissions-in-2019

Inderwildi O, Zhang C, Wang X, Kraft M (2020) The impact of intelligent cyber-physical systems on the decarbonization of energy. Energy Environ Sci. https://doi.org/10.1039/c9ee01919g

Jones P (2016) Transport planning: turning the process on its head. From 'predict and provide' to 'vision and validate'. In: Radical transport conference

Lyons G, Davidson C, Forster T, Sage I, McSaveney J, MacDonald E, Morgan A, Kole A (2014) Future demand - how could or should our transport system evolve in order to support mobility in the future? New Zealand Ministry of Transport, Wellington. https://www.transport.govt.nz/assets/Uploads/Our-Work/Documents/23ed0ae6fc/fd-final-report.pdf

Mulley C (2019) Intelligent mobility and mobility as a service. In: A research agenda for transport policy. Edward Elgar Publishing, pp 187–195. https://doi.org/10.4337/9781788970204.00031. https://www.elgaronline.com/view/edcoll/9781788970198/9781788970198.00031.xml

Nam D, Kim H, Cho J, Jayakrishnan R (2017) A model based on deep learning for predicting travel mode choice. In: Transportation research board 96th annual meeting. https://www.researchgate.net/profile/Daisik_Nam/publication/317913178_A_Model_Based_on_Deep_Learning_for_Predicting_Travel_Mode_Choice/links/59518ed2458515a207f4a01e/A-Model-Based-on-Deep-Learning-for-Predicting-Travel-Mode-Choice.pdf

Naughton B (2020) Chinese industrial policy and the digital silk road: the case of Alibaba in Malaysia. Technical report 1

OECD (2020) Public funding of health care. Technical report, OECD

ORNL (2017) National household travel survey. https://nhts.ornl.gov/vehicle-trips

Ortegon-Sanchez A, Popan C, Tyler N (2017) Car-free initiatives from around the world: concepts for moving to future sustainable mobility. In: Transportation research board 96th annual meeting. Transportation research board. https://s3.amazonaws.com/academia.edu.documents/55849882/CAR-FREE_INITIATIVES_FROM_AROUND_THE_WORLD-_CONCEPTS_FOR_2_MOVING_TO_FUTURE_SUSTAINABLE_MOBILITY.pdf?response-content-disposition=inline%3Bfilename%3DCar-Free_Initiatives_from_Around_the_Wor.pdf&X-

Panter J, Heinen E, Mackett R, Ogilvie D (2016) Impact of new transport infrastructure on walking, cycling, and physical activity. Am J Prev Med 50(2):e45–e53. https://doi.org/10.1016/j.amepre.2015.09.021

Poonawala H, Kolar V, Blandin S, Wynter L, Sahu S (2016) Singapore in motion: insights on public transport service level through farecard and mobile data analytics. In: Proceedings of the ACM SIGKDD international conference on knowledge discovery and data mining, vols 13–17-Augu. Association for computing machinery, New York, NY, USA, pp 589–598. https://doi.org/10.1145/2939672.2939723. https://dl.acm.org/doi/10.1145/2939672.2939723

Rojas-Rueda D, De Nazelle A, Andersen ZJ, Braun-Fahrländer C, Bruha J, Bruhova-Foltynova H, Desqueyroux H, Praznoczy C, Ragettli MS, Tainio M, Nieuwenhuijsen MJ (2016) Health impacts of active transportation in Europe. PLoS ONE 11(3) (2016). https://doi.org/10.1371/journal.pone.0149990

Schrotter G, Hürzeler C (2020) The digital twin of the city of Zurich for urban planning. PFG J Photogram Remote Sens Geoinf Sci 88:3. https://doi.org/10.1007/s41064-020-00092-2. https://doi.org/10.1007/s41064-020-00092-2

Tao F, Qi Q (2019) Make more digital twins. https://doi.org/10.1038/d41586-019-02849-1

Tok A, Ritchie SG (2014) Integration of weigh-in-motion and inductive signature technology for advanced truck monitoring. In: Transportation research board 93rd annual meeting. Transportation research board. https://trid.trb.org/view/1288501https://www.researchgate.net/publication/259822733_Integration_of_Weigh-in-Motion_and_Inductive_Signature_Technology_for_Advanced_Truck_Monitoring

TSC (2018) Blockchain disruption in transport are you decentralised yet? Technical report, Transport Systems Catapult

UNFCCC (2016) Adoption of the Paris agreement, Paris Agreement text English. Technical report, United Nations

UNFCCC (2019) Call by high level climate champion to join the climate ambition alliance at COP25 | UNFCCC. https://unfccc.int/news/call-by-high-level-climate-champion-to-join-the-climate-ambition-alliance-at-cop25

Wadud Z, MacKenzie D, Leiby P (2016) Help or hindrance? The travel, energy and carbon impacts of highly automated vehicles. Transp Res Part A Policy Pract 86:1–18. https://doi.org/10.1016/j.tra.2015.12.001

Wan L, Nochta T, Schooling JM (2019) Developing a city-level digital twin-propositions and a case study. In: International conference on smart infrastructure and construction 2019 (ICSIC). https://doi.org/10.1680/icsic.64669.187. https://doi.org/10.1680/icsic.64669.187PublishedwithpermissionbytheICEundertheCC-BYlicense. http://creativecommons.org/licenses/by/4.0/

Wan L, Yang T, Parlikad A (2019) City-level digital twin experiment for exploring the impacts of digital transformation on journeys to work in the Cambridge sub-region. Technical report, University of Cambridge, Cambridge. https://www.repository.cam.ac.uk/handle/1810/296272. https://www.repository.cam.ac.uk/bitstream/handle/1810/296272/cdbb_ecr_final_report_-_lw_v2_1_-_dr_li_wan_.pdf?sequence=1&isAllowed=y

WHO (2020) WHO/Europe | Transport and health—Physical activity. http://www.euro.who.int/en/health-topics/environment-and-health/Transport-and-health/data-and-statistics/physical-activity2

Chapter 19
Information Driven Energy Markets as Basis for the Energy Transition

Thomas Hamacher

Abstract The energy transformation requires massive investments in renewable energies and a complete transformation of most of the final energy sector. The coupling of the power, heating and traffic sector is very promising. Most of the new renewables produce electricity. This makes electricity the new "primary energy" and most important final energy for all sectors. Second heating and traffic sector offer unique flexibility options, which can balance the intermittent nature of the renewable supply. Still millions of houses need to be refurbished and millions of cars need to be replaced. This transformation process would at best be market-driven to apply cost-optimal solutions and to avoid long-term planning errors. Still, markets lack mainly information on all levels: on the customer side, the supplier side, the regulator side and on the side of official authorities. The creation of a new information platform, which combines various levels of complexity and scales, can overcome this lack of information. Unleashing new market forces promises the advent of a new wave of system transformation.

19.1 Introduction

European, national and regional governments express their wish to reduce the greenhouse gas emissions by 95% compared to emissions in 1990 or become even carbon neutral. The so-called "Green Deal" (COM 2019) of the European Union is looking for the latter goal. A plethora of policy measures was decided to reach this goal. Economists still plea to use only very few simple measures like taxes and emission allowances, but with no overwhelming success. The tendency to make detailed technology policies is imminent. The strong push for alternative drive trains in the mobility sector is a typical example.

The question is, can market forces alone drive such a major transition or is "war economy" necessary to trigger such a major transformation?

T. Hamacher (✉)
Technical University, Munich, Germany
e-mail: thomas.hamacher@tum.de

O. Inderwildi and M. Kraft (eds.), *Intelligent Decarbonisation*, Lecture Notes in Energy 86, https://doi.org/10.1007/978-3-030-86215-2_19

This question will certainly not get a simple answer. Still, it is worth asking it repeatedly and revisiting answers given earlier. At this point, I would like to remind readers of a study issued by Shell, many years ago, which compared a state-driven energy transformation with a market-driven one (Earlier scenarios 2021). The differences were imminent. The scenarios implied that a market-driven change would mainly drive energy efficiency and reduce the overall consumption. The supply side changed only slowly and fossil fuels remained largely in the system. The state-driven scenario developed a massive extension of renewable supplies. Energy efficiency improvements were much slower and consumption did increase. Emission reductions were achieved by a massive increase in renewable energies. The scenarios implied that the regulation scheme would strongly affect the technology choice. The relationship between regulation and technology is well investigated in the history of technology. A meta-discussion about the right regulation is more than timely and necessary.

The goal of this work is to find a compromise between the powerful forces of market economies and the necessary analysis and planning of new advanced infrastructures. While new technologies in telecommunication were the result of liberalized markets, a similar effect was not observed in the energy markets. Economic theory claims that a central planner and decentral market participants will find the same optimum, if a number of requirements are met. A central requirement is symmetry of information. All market participants need to have the same information level. This is certainly quite often not the case and we see strong asymmetries between various market participants.

The central hypothesis of the following work is that the creation of new powerful digital platforms, which combines state of the art data sources with powerful planning algorithms and transparent result presentations, will offer the necessary information symmetry. A constant up-date of data and results is necessary.

The article is organized in the following way.

(1)   The Bavarian energy situation is discussed shortly and major challenges are described
(2)   An analysis framework is presented, which can analyze the power, heating and traffic sector together
(3)   Description of the platform, which publishes these data in various, formats presenting results for nonprofessionals and professionals
(4)   Vision of a new market due to transparency, which drives the energy transformation
(5)   Outlook and conclusion include a remark about the philosophy of science.

The paper is only a very first collection of thoughts and the matter requires much more refined investigations in future. The lack of knowledge, which is imminent, requires more modesty from all actors. Modesty not regarding the overall goals but regarding the knowledge of how best these goals can be reached. The new organically growing knowledge platform replaces static scenarios, which pretend to know the optimal transformation path.

19.2 Power of Information

19.2.1 The Bavarian Energy Situation

The chapter will not deliver a detailed analysis of the Bavarian energy system but just a few hints about the hurdles and problems to transform the system to a low carbon system.

According to the Bavarian Ministry of Economic Affairs, Regional Development and Energy Bavaria consumed 540 TWh of primary energy and 395 TWh of final energy in the year 2017 (Bayerisches Landesamt für Statistik 2017).

In the year 2017 Bavaria consumed 78 TWh of power, just 20% of the overall final demand. Nuclear power plants produced 31 TWh and renewable technologies 37 TWh (Bayerisches Landesamt für Statistik 2017). The CO_2-Emissions in the Bavarian power system are rather low, because nuclear power and hydropower played quite a crucial role. The decision to stop nuclear until 2022 is a major thread for the emission balance. Many studies and investigations proposed options to replace the nuclear power plants. Three major options are favored: the extension of power lines to Northern Germany, the construction of gas power plants and increase of renewable production. (1) The construction of new power lines faces major resistance from various groups. Despite the construction of new lines being decided, still the construction process is slow. (2) The Bavarian government pushed for more gas power plants. Still, it is impossible to find investors under current market conditions. Bavaria hosts some very modern gas power plants, which are hardly used in the market. Electricity from renewable sources and from coal and nuclear power plants are more competitive. The use of gas would certainly increase greenhouse gas emissions. (3) PV plays a central role in Bavaria and will certainly be extended in future. Still, PV alone is not able to supply a power system in a Northern Country. The increase of wind energy is much slower. This is due to an overall resentment in the public. A second reason for the slow increase in wind power is the so-called 10 h rule. The rule requires that wind turbines need to be more than ten times their height away from settlements. In any case, Bavaria needs to extend renewable production massively not only to cover the power consumption and replace missing nuclear power but to deliver power to the other sectors discussed below. However, whatever happens in the short run, strong connections to other parts of Germany and neighboring countries will keep the power system running.

The heating sector is still dominated by conventional technologies especially gas and oil boilers. In households alone, both technologies supply up to 63 TWh of heat in 2017. Renewable energies play a minor role in the heating sector jet. Some parts of Bavaria have excellent sites for deep geothermal heat and some have substantial amounts of biomass. The building sector is diverse with very rural parts at one end of the scale and metropolitan areas like Munich and Nuremberg at the other side. The refurbishment rate did also not accelerate recently. A clear idea to transform the sector is missing.

Cars and trucks consumed 110 TWh of gasoline and diesel in 2017. Mobility plays a crucial role with many people living in the countryside and working in the cities, with many leisure-time attractions within a ride of a few hours. The general policy is to increase e-mobility. Hydrogen is also discussed as an option for trucks and trains.

The challenges are enormous. Renewable production needs to increase dramatically. The future use of hydropower and biomass is not expected to increase. PV and wind need to carry the main part of the supply. New technologies like geothermal for the heat supply need to be developed in the coming years. However, the increase of production is only one side of the coin. On the other side, the final energy sector needs to be transformed to electricity as the final energy carrier. Heating systems in millions of buildings and millions of cars need to be replaced.

19.2.2 New Data and Analysis Frameworks

Let us start the discussion with a simple example. A house owner in the Bavarian Forest needs to change the heating system in his single-family house with a living space of 200 m^2. The house is heated by a 25 years old central oil boiler. A first energetic refurbishment of windows and walls reduced the annual heat consumption to 120 kWh/m^2 excluding hot water. What should he do next? Friends with a new house bought a heat pump; others added a solar collector on the roof and reduced the oil consumption for heating by 40%.

A company in a middle-sized Bavarian city has more than 1000 employees. Most of them commute by car. Average daily distance is well below 100 km/d. Many employees consider buying an electric car but like to have electric charging opportunities at their workplace. What should the company do?

Both examples are rather normal situations in a rapidly changing energy environment. At the time being it would be rather difficult to give sound advice for both of them right on the spot.

Certainly, we have developed many tools in the past, which could deliver answers. But these models are either much too coarse and will only give general answers, which are certainly unable to justify a major investment or they are very specific and lack a connection to general developments and might miss out opportunities or problems. The new data and model environments need to combine these scales. Data sources and models need to qualify according to some standards. In an ideal world, public authorities would mainly provide data and models. Scientists would certify both. Algorithms and programs are open-source (Fig. 19.1).

In a first attempt, the models to describe the heating and the power sector need to be combined. The heating sector requires finally data for each house and needs to be modeled at this level. Chimney sweepers have at least detailed information for all buildings with fossil heating systems, but these data are not available. Therefore, we propose the development of synthetic data. Classical information combined with various sources and machine learning techniques will be used to develop an idea

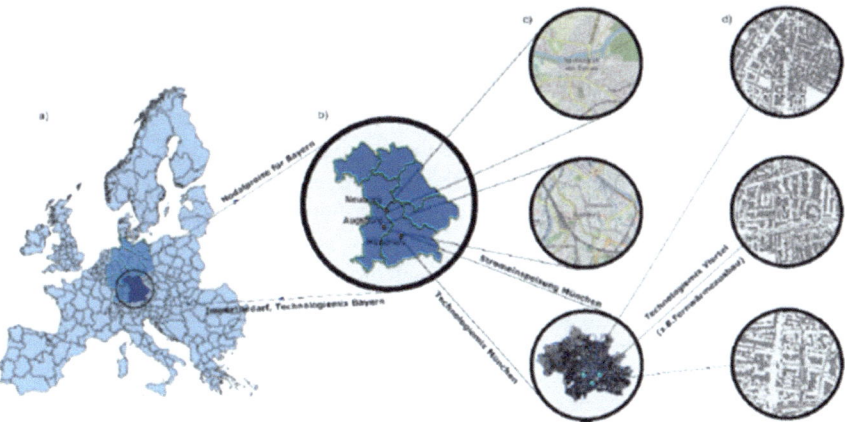

Fig. 19.1 Data and models need to cover all scales from Europe down to quarters and finally houses and companies. The connection between all levels is crucial

how the supply structure and the refurbishment state of a certain region is. The data can at all stages be replaced by real data and improve then the analysis. The concept is to start with a high level of detail right from the beginning of the analysis, but the uncertainty is high at the beginning and will then gradually improve with more analysis steps done. The development of synthetic data is an art in itself. The merits are obvious since it offers the opportunity to start the analysis right away with detailed data and some idea about the uncertainty. In the process more and more of the synthetic data will be replaced by real data.

The next challenge is to combine the different scales and get the interaction right. Again, let us demonstrate this with an example. If the power sector goes ahead with a fast implementation of wind on- and off-shore turbines and the power lines between north and south Germany are strongly enhanced even beyond the current planning. Then the wide intense application of heat pumps even in Bavaria will make sense. Wind energy produces more power during winter, which is well correlated with the increased heating demand in winter times. A European power model can certainly deliver here the necessary information. However, the development might not come as expected and therefore it might be necessary to find indicators, which tell us to what extent usage of heat pumps still makes sense under current conditions. Model and development need now to be closely connected and it is extremely necessary to derive indicators, which allow us to estimate if a development will happen as expected or might fail. The new power lines serve as an example. Design, planning, licensing and final construction take time. At least simple estimates exist about the duration of each step. It is now possible to backtrack and identify if model results and real development still coincide. In case the coincidence is no longer guaranteed, the model needs to include the delay and find alternatives. For our example, it could mean that from a certain point in time on the installation of heat pumps in Bavaria is reduced and alternative techniques are favored. In modern language, we would consider the model to be a digital twin, which grows and shrinks with the real system.

The same is true for small-scale developments. If—as we observe today—most house owners stick to conventional heating technologies and buy new oil and gas burners it would be naive not to assume that these boilers are in ten and most likely twenty years still in operation. Again, the model needs to monitor the sales numbers of these technologies and it needs to make sure that they remain in the model for a certain time.

The combination of models at different model scales is done by decomposition methods. These methods are well established in optimization theory. Still, the application to our case needs to be improved.

19.2.3 The New Energy Information Platform

The new data and analysis tools need to be organized on a new platform, which is certainly overlooked by public authorities, but which is mainly a matchmaking tool. I would like to start again with an example. Assume again a village in the Bavarian Forest. A small quarter of the village was built in the early nineties and the heating systems requires in most of the houses a refurbishment. Homeowners will certainly consider the classical options and might play with some additions like solar thermal panels on the roofs. Still, most ideas are rather conservative. Assume we have a young start-up company, which specialized in small district heating networks. The grid connects only a couple of houses. The combination makes it possible to use a portfolio of different heating technologies. The new platform had of course delivered a first analysis of the village. Estimates about heating technologies make a more detailed analysis possible. The new company performs the analysis for the mentioned houses and concludes that an investment might be viable. They contact house owners and the local authorities. This would certainly make business processes much better and it would especially allow all kinds of businesses to look for opportunities. Already we have companies, which have many of the data, which would be necessary to highlight the mentioned opportunities, but they would certainly not be willing to share the information with others. Once new investments are done, they are documented in the platform. This again improves the overall modelling detail.

A central question is who will host and maintain such a platform. Bavaria has started the first step in this direction with the Energy Atlas of Bavaria. Still, the platform discussed here is still quite some way to go. Data privacy issues will play quite a role.

19.2.4   Vision of a New Market Due to Transparency

New technologies make more transparency possible even if the data is not shared openly in the first place. A smart combination of publicly available data, remote sensing data and general information on geography and climatic conditions will allow sound estimates about building conditions and utilized technologies. Extra information from informal sources will certainly complement the picture. While this vision sounds rather promising, in the first place it might also turn out as a nightmare with numerous marketing campaigns, which flood over citizens, companies and administrations. The task of the platform is to challenge unfounded marketing claims and keep customers well informed at low cost in time and money.

Internet platforms are used intensively today to prepare purchase decisions by comparing prices and services. For example, the platform operated by check24 is used widely in Germany.

The newly gained information symmetry will boost then new market developments. The return of market-driven developments is certainly a new phase in the energy transition. Feed-in tariffs and purchasing premiums have for long been the tools of choice. The return of markets would certainly help a faster transition but require trust and honesty from all market participants and technology agnostic policies.

19.3   Conclusions and Outlook

Many people still believe that more regulation, more state programs and fewer market-driven efforts will promote the energy transformation. The first error in most of these concepts is that we know what we need to avoid doing, but that we usually do not know what to do. The support for certain technologies comes and goes in waves. The government is certainly unable to make long-term predictions and so is science. Socio-economic uncertainties and technological developments make long-term forecasts impossible. This is certainly accepted by most participants in theory while the behavior indicates the opposite. Therefore, it is essential that the system stays open for new developments. Investments in new infrastructure create a certain lock-in, but this should be kept within manageable limits.

The scientific community needs to develop a much better understanding about the limits of pictures of the future. A philosophy of energy modelling needs to complement the modelling work. The limit between sound scientific results and necessary political decisions needs to be clear. In times of major uncertainty, science needs to be honest and modest. Policy needs to make decisions nonetheless. However, it is of course much better if it is known that a decision might be wrong and needs to be open for correction instead of pretending to be a scientific necessity. The incorporation of uncertainty in the regulation is necessary. Open data platforms which daily measure the performance of a policy will help to redirect decisions and steer the overall system to a better end.

The new energy information platform combining data, analysis tools and analysis results will allow the transformation to be open for new developments. Up-dates on relevant developments and the steady interaction between all scales of the system make a market-driven energy transformation possible.

References

Bayerisches Landesamt für Statistik—Energiebilanz (2017)
European Commission (2019) The European Green Deal COM(2019) 640 final. Brussels, 11 Dec 2019
Earlier scenarios. https://www.shell.com/energy-and-innovation/the-energy-future/scenarios/new-lenses-on-the-future/earlier-scenarios.html. Accessed 17 Jan 2021

Part IV
Legal & Governance

Chapter 20
The Artificial Intelligence Governance Gap: A Barrier to Intelligent Decarbonization

Thorsten Jelinek

Abstract The notion of intelligent decarbonization (IDC) suggests that artificial intelligence (AI) is a significant part of the solution to climate change, thus resolving the contradiction between economic growth and sustainability preventing the transition towards a low-carbon economy. While technology plays a vital role in combating climate change, over-reliance on technology and AI, in particular, presents a tremendous risk. AI is not only attributed as a unique opportunity for efficiency with the potential to solve some of humanity's greatest challenges, but also as a source of unprecedented cyber-physical threats and structural imbalances. Due to existing structural forces and technology determinism, those risks will be sustained in the course of AI's development and adaptation. Thus, in addition to the existing climate action gap, IDC inherits AI's emerging governance gap, which presents new barriers to the goal of decarbonization. To address those barriers effectively, as highlighted in the outlook of this chapter, IDC must balance the technological and political dimensions of IDC governance. However, only the latter has the potential to counter technology determinism and intervention on a structural level. A global IDC coordination mechanism guided by human- and environment-centric principles is needed to support the development of local IDC governance networks.

20.1 Introduction

The notion of intelligent decarbonization (IDC) suggests that technology and, in particular, artificial intelligence (AI), is a significant part of the solution to climate change. AI has been widely attributed with the potential to help realize the United Nations Sustainable Development Goals (SDGs), of which decarbonization is a key factor (Vinuesa et al. 2020). On a less aggregated level, an ambitious IDC agenda is emerging (Rolnick et al. 2019). Accordingly, AI in combination with cyber-physical systems (CPSs) is thought to contribute to the reduction, elimination, and removal

T. Jelinek (✉)
Taihe Institute, 23/F, ShunMaiJinZuan Plaza, A-52 Southern East Third Ring Road. Chaoyang District, Beijing 100022, China
e-mail: thorstenjelinek@taiheglobal.org

© Springer Nature Switzerland AG 2022   183
O. Inderwildi and M. Kraft (eds.), *Intelligent Decarbonisation*, Lecture Notes in Energy 86, https://doi.org/10.1007/978-3-030-86215-2_20

of carbon dioxide emissions; the mitigation of unavoidable consequences, including planning for resilience and disaster recovery related to the impact of storms, droughts, fires, and floods on the environment, economy and society; and the optimization of decision making related to markets, policies, and consumer behavior. With the further advancement and adaptation of AI, IDC is going to target all major sectors of greenhouse gas emissions (GHG), including energy provision, industrial production, transport management, urban organization, and land use.

At this early stage of development, however, IDC largely remains speculative. Thus, to avoid presenting this emerging domain of climate action as a *digital utopia* that will resolve today's contradiction between economic growth and sustainability, proponents of IDC also warn against its technological limitations and broader nontechnological factors that could prohibit the realization of the SDGs and represent barriers to the important goal of decarbonization (Rolnick et al. 2019; Vinuesa et al. 2020). The technological limitations are related but not limited to the availability and accessibility of abundant, high-quality data as well as of CPSs; the potential lack of adaptability of AI solutions across sectors and regions; unintended reverse effects of optimizations that could strengthen rather than change the existing carbon-based energy system; and the immense energy consumption of AI itself, which is counterproductive if energy sources remain based on fossil fuels. In contrast, nontechnical factors refer to potentially negative impacts on the environment, economy, and society due to the widespread diffusion and application of AI. While the resolution of technological limitations is largely a matter of science and engineering, the structural factors refer to existing historical forces as well as an overreliance on technology itself, which becomes reinforced through AI. As discussed in the following, AI has been cautioned against as a source of unprecedented threats and imbalances. The nature of those risks is explained in terms of AI's *dual-use character* (Brundage et al. 2018) and the more general impact of *technology determinism* (Brockelman et al. 2009; Feenberg 2001; Wyatt 2008). Both represent tremendous obstacles to the implementation and governance of responsible AI. IDC not only faces the AI's governance gap but also the climate action gap, both of which risk undermining the impact of IDC and fight against climate change. As proposed in the outlook section, what is needed is the development of local IDC governance networks and a global IDC coordination mechanism to tackle those obstacles most effectively.

20.2   AI's Dual Use as Constitutive Void

Like climate change, AI is seen as one of humanity's global challenges. Its dual-use character suggests that the application of AI can result in good or in harm. On the upside, AI is considered a new source of innovation, economic growth, and competitiveness, offering improved efficiency across entire economies. AI also has the potential to resolve some of the most urgent global challenges, including the fight against climate change. Governments have rushed into executing their AI strategies to capture the potential upside deemed necessary to sustain the future competitiveness of their countries (Dutton 2019). On the downside, AI's disparate technologies (Corea 2018) have been linked to various risk scenarios across applications, sectors, and geographies. Those scenarios can be clustered into two groups based on immediacy and scope of the risk: (i) direct cyber-physical threats and (ii) longer-term structural imbalances (cf. Brundage et al. 2018; Zuboff 2018).

Firstly, AI alters the landscape of direct cyber-physical threats as the result of adversarial use. This entails an expansion of existing threats, more effective and targeted threats, and the emergence of entirely new types of cyber-physical threats. In addition to such intended attacks, there will be unintended and unpredictable accidents due to engineering bottlenecks, which will also be the target of intentional exploitation. AI, together with ubiquitous digitalization, will intensify the issue of cyber security; for many governments, this is already a matter of national security and protecting critical infrastructure, including the energy sector (Viganò et al. 2020).

Secondly, the structural imbalances have longer-term consequences and are more difficult to determine and anticipate, but their impact is expected to be much more widespread and pervasive. They broadly refer to AI's impact on human relations and dignity by constraining both power and opportunity. AI is thought to change all dimensions of human affairs, including the economy, society, politics, and international relations. Economically, vast labor displacement, underemployment, and de-skilling are likely outcomes as well as an unequal concentration of AI capabilities and benefits. For societies, increasing lack of dignity, privacy, and meaning will threaten personal well-being and social cohesion. Politically, AI increases the structural risk of shifting power balances between the state, economy, and society by limiting the space for autonomy. Internationally, fierce global competition over technology and AI leadership risks further fragmenting existing international relations and regime complexes. The proliferation and accessibility of offensive cyber capabilities, particularly lethal autonomous weapons, increases the risk of ongoing asymmetric conflicts and "hyper war" (Husain 2018). Finally, AI is also linked to the distant scenario of a trans- and posthumanist era, which could pose an existential threat to humanity altogether (Ferrando 2013).

20.2.1 Repeated Through Technology Determinism

AI's dual-use character doesn't provide a simple choice between good or harm, as the good-or-harm dichotomy manifests as a constitutive void that is already occupied by broader historical developments. In spite of potentially unprecedented risks, AI will still lead to widespread adaptation, not only due to its vast efficiency potential but also to a deeply rooted belief in science and technology (Brockelman et al. 2009; Feenberg 2001; Johnston 2018). Science, as the basis of technology and with the expansion of capitalism since the first industrial revolution, has determined our Anthropocene age and created a path dependency from which history cannot simply depart. In particular, science is "the 'Real' of our historical moment, that which 'remains the same' in all possible […] symbolic universes" (Zizek 1997, p. 38).

Although science is a human artefact, it is not a mere social construction or an expression of the interest of a certain group. It cannot easily be replaced by another cognitive dialectic based on hypothesis and verification. As a result, modern life is intrinsically related to continued objectification in terms of mathematical laws. Scientific drive and capitalist productivity constitute two fundamental ethics that explain why AI's dual-use dichotomy and the void it creates are carried forward in history. They explain the overreliance on technology for solving human problems, even though such technology determinism is part of the problem as well. Equivalently, though science has been the primary force behind human emancipation, science also has a repressive dimension, as its laws limit the space of autonomy and subjectivity. Hence, the fear of dystopian AI lies in its potentially all-encompassing objectification of human experiences. The access to all factual knowledge might not only threaten jobs, undermine social cohesion, and shift power relations and equity, but also deprive humans of the subjectivity through which we experience our autonomy and freedom (Zizek 2012, p. 148). As a consequence, social and environmental good are not an intrinsic outcome of techno-scientific and capitalist economies but require political interventions and decisions concerning ethics and governance.

20.2.2 Obstacles to AI Ethics and Governance

The rapidly emerging and increasingly materializing landscape of AI risk and dystopian scenarios have sparked international debate about the ethics and governance of AI. This debate has resulted in the definition of numerous normative principles and frameworks worldwide (Jobin et al. 2019; Zeng et al. 2018) and triggered the rise of a comprehensive AI governance research agenda (Dafoe 2018). Despite differences in selecting and interpreting certain normative principles, those frameworks commonly emphasize that AI should be developed and used responsibly for the greater good of humanity. In particular, those frameworks commonly stress that AI should be secure, safe, explainable, fair, and reliable, and that its benefits should primarily serve society. Given the all-encompassing impact of AI, the private and public sector are urged to use these principles to guide various aspects of AI development and use, not only based on governmental policies but also on industry standards, laboratory practices and procedures, and engineering solutions, and consumers should be aware of AI's risks.

Such normative framing of the future use of AI conversely underpins the widespread concern that today's governance—based on its underlying principles from the outset—is insufficiently guided and structured to prevent and mitigate harm effectively (Wallach and Marchant 2019). The emphasis on responsible AI highlights a mismatch or insufficient overlap between those new external intentions (principles of responsible AI) and existing internalized actions (actual ethics) as well as between existing and future instructions for collective action (governance). While those AI principles have been defined fairly rapidly, respective governance approaches are only slowly emerging. The reason for such a prolonged gap is that AI governance operates precisely upon the constitutive void of AI's dual-use dichotomy. The uncertainty related to responsible AI governance is certainly owed to the rapidly emerging yet still speculative nature of AI. Existing regulations and regulatory approaches do not match such complexity, nor can they keep up with the speed of AI's advancement and adaptation. Initially, this reinforces awareness of and increases the governance gap. However, the lack of a simple choice between good and harm and the exposure to existing realities will prevent the closure of the governance gap altogether. Thus, AI governance will continue facing the absence of a plain choice but can only aspire for the good to sufficiently outweigh the harm. More concretely, there are at least two realities or structural forces that occupy the void, undermine the governance of AI, and risk turning responsible principles into a mere ideology of the Anthropocene. Those core forces also undermine climate action and have turned humanity into a disruptive "geological factor" (Wark 2015).

First, AI governance is influenced by fierce competition over global AI leadership, which has become reinforced through various national AI strategies. Competition fosters innovation but not necessarily progress in terms of responsible AI. Fierce competition tends to compromise responsibility and lead to a concentration of AI resources and power imbalances. Ultimately, techno-scientific capitalism lacks the ethics of social and environmental good, primarily thriving on the continuous extraction and utilization of resources and self-interest. This will mainly determine the development and use of AI if those dominant ethics remain unchecked and unchanged. Second, cultural differences and competing political interests and government systems lead to conflicting normative frameworks and regulations. They increase tension between state actors and further undermine much-needed collaboration and coordination. Those differences and tensions are perpetuated through rising nationalism, protectionism, and populism and a heightened distrust in multilateralism, public institutions, and the private sector (OECD 2019).

20.3   Conclusions

These broader developments and dependencies suggest that there is no easy reconciliation between both sides of the AI governance gap. Since the beginning of the 2000s, the world has entered a period of constant economic, geopolitical, and health crises, and this seemingly insurmountable gap between a digital utopia and dystopia is not balanced but skewed towards the latter (Jelinek 2020; Jelinek et al. 2020). The seemingly insurmountable gap that AI governance increasingly faces has already affected climate action for many years. Twenty-five years passed between the first report from the Intergovernmental Panel on Climate Change (IPCC) in 1990 and the adaptation of the Paris Agreement in 2015, and it took another three years for the international community to agree upon common rules for the adaptation of the Paris Agreement. Despite such efforts and a strong public awareness, greenhouse gas emissions have remained on the rise year after year, and it has become less likely that the average warming will be limited to 1.5 °C. "Brown recoveries" (Hepburn et al. 2020) are more likely at times of crises, as they more quickly recover the existing system, through which wealth and economic behavior remains tightly linked to linear value chains and excessive fossil fuel consumption, being reproduced through sectoral and national interests. IDC inherits both the AI governance gap and climate action gap and risks the ideological trap of technological solutionism (Johnston 2018; Wyatt 2008).

20.4   Outlook

Paradoxically, technology alone will not bridge the economy and sustainability, but it must be a core dimension of climate action. Thus, for IDC to effectively address the structural forces outlined above and become a coordinated and impactful force of climate action, it needs to be established as a "non-state actor" (Raustiala 2012) adopting both *network governance* and *meta-governance* approaches that adapt and promote common environment- and human-centric AI principles. Governance is part of the general political system, and while IDC focuses on the technological dimension of decarbonization, it is the political dimension that counters the risk of technological solutionism (Johnston 2018) while fostering collaboration and coordination on an international level.

In particular, an IDC network governance approach is a bottom-up, self-organized, and polycentric approach for developing and fostering an entire ecosystem of IDC skills, technologies, solutions, and oversight practices targeted at AI-enabled cyber-physical systems (cf. Shackelford 2020; Wallach and Marchant 2019). As the development and diffusion of IDC solutions cannot rely solely on market dynamics, they must also be established and managed as "commons" (Ostrom 1990). Public–private partnerships should be forged to promote such public good and act upon a vision for a "caring economy" (Snower et al. 2017). For the IDC ecosystem to minimize its risks while maximizing the benefits for society and the environment, it must continuously identify and resolve technology limitations as well as influence related policies, industry standards, laboratory practices, and engineering solutions. The advantage of network-driven governance arrangements (Cihon et al. 2020; Shackelford 2020) is their efficiency in identifying the wide range of technological uncertainties, policy issues, and innovative solutions to be adjusted to local requirements.

However, the emerging IDC ecosystem (which will be comprised increasingly of specific technology solutions as well as diverse networks of local private, public, and civic actors of IDC governance) needs structural support and systemic integration to strengthen those IDC networks and increase their efficiency and impact (cf. Buchanan and Keohane 2006; Cihon et al. 2020). Such support can mainly be achieved through more centralization or meta-governance (Jessop 2011). Here, meta-governance is a coordinating function intended to establish and strengthen institutional linkages within and between the climate change (Keohane and Victor 2010) and cyber regime complexes (Nye 2014) mainly comprised of governmental and intergovernmental organizations. Such IDC meta-governance mechanism could be part of an existing multilateral or government formation, like the Group of Twenty (G20) giving its influence on international policy and representation of 85% of global GDP (Jelinek et al. 2020). The strategy would be to institutionalize the IDC as meta-governance function with the intent to synchronize, integrate, and delegate responsibilities and decision-making between competencies and organizations as well as sharing oversight outcomes and catalyze effective policy instruments that have already been promulgated or proposed. The objective of IDC meta-governance is to help implement a global agenda for IDC and achieve a higher degree of integration and coherence of both regime complexes, which would help to counter their fragmentation and simultaneously strengthen local IDC governance networks.

The basis for effective IDC governance and meta-governance is strategic foresight (Buchanan and Keohane 2006). To reduce the conflict between sustainability and growth, IDC foresight must achieve an overview of complementary and conflicting development objectives and regulatory policies as well as instruments concerning IDC and AI risks. Hence, a policy coherence approach (cf. Miola et al. 2019) is instrumental in monitoring the development and application of IDC-related regulations and, especially, in analyzing how they fit together, where they agree, and what gaps and conflicts need to be addressed. As a starting point, the SDGs provide the most comprehensive and integrated development framework of objectives, targets, and topics requiring a coordinated approach from several domains. The development of such foresight can already build upon existing research into the role of AI in achieving the SDGs and tackling climate change (Rolnick et al. 2019; Vinuesa et al. 2020). Both analyses need to be extended and mapped with a detailed review of the AI threats and risks landscape to achieve an IDC program that most effectively resists becoming an ideology of our Anthropocene age and fails to improve the impact of climate action significantly.

References

Brockelman TP (2009) Zizek and Heidegger: the question concerning techno-capitalism. Bloomsbury Publishing, New York

Brundage M, Avin S, Clark J et al (2018) The malicious use of artificial intelligence: forecasting, prevention and mitigation. Technical Report, arXiv:1802.07228v1

Buchanan A, Keohane RO (2006) The legitimacy of global governance institutions. Ethics Int Aff. https://doi.org/10.1111/j.1747-7093.2006.00043.x

Cihon P, Maas M, Kemp L (2020) Should artificial intelligence governance be centralized? Design Lessons from History, arXiv:2001.03573v1

Corea F (2018) AI knowledge map: how to classify AI technologies. https://link.medium.com/0f4 BD3u754. Accessed 29 Aug 2019

Dafoe A (2018) AI governance: a research agenda: working document v1.0. Governance of AI Program Future of Humanity Institute University of Oxford, Oxford

Dutton T (2019) An overview of national AI strategies. https://link.medium.com/MNCJQvLBk8. Accessed 20 Feb 2019

Feenberg A (2001) Transforming technology: a critical theory revisited. Oxford University Press, Oxford

Ferrando F (2013) Posthumanism, transhumanism, antihumanism, metahumanism, and new materialisms differences and relations. Existenz 8(2):26–32

Hepburn C, O'Callaghan B, Stern N, Stiglitz J, Zenghelis D (2020) Will COVID-19 Fiscal recovery packages accelerate or retard progress on climate change? Oxford Review of Economic Policy, Oxford

Husain A (2018) Hyperwar: conflict and competition in the AI century. SparkCognition Press, Austin

Jelinek T, Wallach W, Kerimi D (2020) Coordinating committee for the governance of artificial intelligence, The G20 Insights Platform, Global Solutions Initiative. AI Ethics. https://doi.org/10.1007/s43681-020-00019-y

Jelinek T (2020) The future rulers? In: Billows W, Körber S (eds) Reset Europe. Culture Report EUNIC Yearbook 2020. European National Institutes for Culture (EUNIC) and Institut für Auslandsbeziehungen (ifa), Steidl, Göttingen, pp 244–252

Jessop B (2011) Metagovernance. In: Bevir M (ed) The SAGE handbook of governance. SAGE, London pp 106–123

Jobin A, Ienca M, Vayena E (2019) The global landscape of AI ethics guidelines. arXiv:1906.11668

Johnston SF (2018) The technological fix as social cure-all: origins and implications. IEEE Technol Soc 37(1):47–54

Keohane RO, Victor DG (2010) The regime complex for climate change. Discussion paper 10–33. The Harvard Project on Climate Agreements, Belfer Center

Miola A, Borchardt S et al (2019) Interlinkages and policy coherence for the sustainable development goals implementation: an operational method to identify trade-offs and co-benefits in a systemic way, EUR 29646EN. Publications Office of the European Union, Luxembourg

Nye JS (2014) The regime complex for managing global cyber activities. Belfer Center for Science and International Affairs, Harvard Kennedy School, Cambridge

OECD, 2019.OECD (2019) OECD business and finance outlook 2019: strengthening trust in business. OECD Publishing, Paris. https://doi.org/10.1787/af784794-en

Ostrom E (1990) Governing the commons: the evolution of institutions for collective action. Cambridge University Press, Cambridge UK

Raustiala K (2012) Non-state actors in the global climate regime, UCLA Law School. Public law and legal theory research paper series. Research Paper no. 7–29

Rolnick D, Donti PL et al (2019) Tackling climate change with machine learning. arXiv:1906.054 33v2

Shackelford SJ (forthcoming 2020) The future of frontiers, Lewis & Clark law review. Kelley School of Business Research Paper No. 19–12

Snower D, Chierchia G, Lesemann P et al (2017) Caring cooperators and powerful punishers: differential effects of induced care and power motivation on different types of economic decision making. Sci Rep 7:11068. https://doi.org/10.1038/s41598-017-11580-8

Viganò E, Loi M, Yaghmaei E (2020) Cybersecurity of critical infrastructure In: Christen M, Gordijn B, Loi M (eds) The ethics of cybersecurity. The international library of ethics, law and technology, vol 21, pp 456–466

Vinuesa R, Azizpour H et al (2020) The role of artificial intelligence in achieving the Sustainable Development Goals. Nat Commun 11(233):1–10

Wallach W, Marchant GE (2019) Toward the agile and comprehensive international governance of AI and robotics. Proc IEEE 107:505–508

Wark M (2015) Molecular red: theory for the Anthropocene. Verso, London

Wyatt S (2008) Technological determinism is dead; long live technological determinism. In: Hackett E, Amsterdamska O, Lynch M, Wajcman J (eds) Handbook of science and technology studies. MIT Press, Cambridge, pp 165–180

Zeng Y, Lu E, Huangfu C (2018) Linking artificial intelligence principles. arXiv:1812.04814

Zizek S (1997) The plague of fantasies. Verso Books, London
Zizek S (2012) Less than nothing: Hegel and the shadow of dialectical materialism. Verso Books, London
Zuboff S (2018) The age of surveillance capitalism: the fight for a human future at the new frontier of power. Profile Books, London

Chapter 21
Insights: Cyber Security

Alon Cliff-Tavor

21.1   What Are the High-Level Trends You Identify?

It is no secret that digitalisation has had and will continue to have profound effects on most, if not all, industries. Dramatic developments in areas like computing power availability, storage cost reduction (leading to proliferation of data available for use and analysis), ultra-connectivity and the widespread distribution of smart and connected devices, alongside shifts in business and industry perceptions of how these developments could be harnessed, continue to send shockwaves across all the industries we serve.

Cyber and information security has been evolving rapidly and is also a trend, and a critical enabler of digitalisation. 80% of firms recently surveyed reported an increase in cyber-attack. Global losses from cybercrime skyrocketed to nearly US$1 trillion in 2020.[1] Billions of records are lost in both small-and-targeted attacks as well as in large-scale ones.

While in the past we saw motivations for cyber-crime being a mix of financial theft, "hacktivism" and a sport-like challenge attracting mostly young hacking enthusiasts, we now see cyber-attacks being predominantly motivated by financial gain (more than 70%), and state-sponsored (or state-related) espionage (~25%[2]) being responsible for almost all the rest. We see strong, powerful and well-funded organised crime investing substantial amounts in building offensive cyber capabilities, while we continue to hear about nation states weaponising cyber. As a result, the global annual spend on security is expected to grow to more than US$170 billion by next year.[3] Needless

[1] McAfee, The Hidden Cost of Cyber Crime.
[2] Verizon, 2020 Data Breach Investigations Report.
[3] Gartner, Forecast Analysis: Information Security, Worldwide Update.

A. Cliff-Tavor (✉)
Oliver Wyman, Singapore, Singapore
e-mail: alon.clifftavor@oliverwyman.com

© Springer Nature Switzerland AG 2022
O. Inderwildi and M. Kraft (eds.), *Intelligent Decarbonisation*, Lecture Notes in Energy 86, https://doi.org/10.1007/978-3-030-86215-2_21

to say, if corporates, governments and individuals can't trust that the data kept by their providers and partners is safe, they will be far less likely to engage in digital activities related to these providers and partners.

21.2 How Many of Your Digitalisation Clients Are Working in the Sustainability Sector? Do You See Clients Investing in Digitalisation Thanks to Sustainability Motivations?

Although we have no sustainability sector clients working with us on digital transformation, there are several with a focus on multiple digitalisation initiatives that will have a substantial impact on sustainability.

For example, we are working with several clients to help their organisations move from captive, private data centres to public cloud service providers. This alone, when examined at a large scale, has a profound environmental impact compared to the alternative. Cloud providers today are using a far smaller number of servers (and other resources) and are managing their data centres in far more efficient ways compared to private enterprise data centres. This translates[4] into far less hardware having to be manufactured and transported around the world and much lower energy consumption. We see more clients across sectors considering the environmental impact of their transformation and counting their technology investments towards meeting their decarbonisation objectives.

21.3 What Do You See is the Value and Impact of Blockchain?

It has been said many times that blockchain technology is a wonderful solution in search of a real, suitable problem. I believe that such problems and use cases are emerging and there are hundreds of ventures and enthusiasts aiming to leverage this technology to solve different problems and opportunities. As of now, I do not believe that blockchain has had a dramatic effect on any particular industry or segment. I do, however expect blockchain to be adopted for future use cases such as supply chain management and security, payments, fraud prevention and registry management.

In the context of decarbonisation and sustainability, I expect that blockchain could help to validate and verify the authenticity of carbon asset management for applications such as carbon trading and fraud prevention.

---

[4] Mission, an Amazon Web Services managed services partner firm, estimates that following a cloud transition an enterprise will consume 77% fewer servers, 84% less power and reduce carbon emissions by 88%.

21.4 How Do You See Smart-Everything Evolving in the Future? Any Threats to That?

I absolutely expect a strong continuation and dramatic increase in the "smart-anything" trend. The proliferation of data, computing power, connectivity and analytics techniques provides endless possibilities for impacting industries, helping to improve the environmental and economic efficiency of existing processes, products and services, and making new technologies, products and services economically sustainable.

I do see some obstacles, challenges and risks to these trends. The main one might be related to information security and privacy protection: with the entrenched and deepening surveillance that smart city capabilities would require, I expect substantial public privacy concerns. Machine learning and artificial intelligence are just as susceptible to cyber-attacks and other forms of manipulation, if not more. Regardless of a threat actor's identity or motivation, expect manipulative efforts targeting AI applications to increase. Effectively securing the infrastructure, systems and data assets related to smart cities is a necessity and unless security can be assured, I expect progress to miss its full potential.

Alon Cliff-Tavor is a Partner at Oliver Wyman, based in the firm's Singapore office. He is a member of the OW Digital and Financial Services practices and has substantial experience in IT strategy, IT development, security, operational excellence, business transformation and change management. His current client work focuses on digital transformation, technology modernisation, cyber, information security and technology risk management. Prior to joining Oliver Wyman, Alon worked for several institutions across Asia, delivering large-scale transformational programmes to front, middle and back-office functions, usually involving substantial organisational and process re-design, enabled through technology. He studied Law and Economics and holds an MBA from INSEAD Business School, France.

Chapter 22
Responsibility for Artificial Intelligence

Michael Neupert

Abstract Intelligent decarbonisation utilizes artificial intelligence. Apart from technical issues, a discussion has begun about who is responsible for potential mistakes made by this technology. To order the basic structures, a loan from legal theory can help separate autonomous machines from complex automatons and show in which direction legal responsibility will evolve. Even if many detailed questions will still have to be clarified, a basic tendency, which is relevant for strategic planning, shows up in this way.

22.1 Introduction

Law sets the framework for social developments and is one of the most important instruments for implementing political decisions. With regard to decarbonisation, a number of legal instruments come to mind,[1] but they are not the subject presently discussed. Law also plays an important role in regulating technology and, in particular, its advances, such as the use of artificial intelligence. A fundamental debate has arisen about how legal systems should deal with the topic of responsibility once intelligent machines make decisions on their own, even if only to a certain extent.

Responsibility plays a key role in the legal context of intelligent decarbonisation. Not only will the regulator consider whether specific rules are necessary, but also whether distribution of risk will foreseeably play a role in the contractual relationships, according to which businesses structure their operations in unchartered waters. This raises fundamental questions: Is there a vacuum of responsibility? Is law fundamentally overwhelmed by the novel technology of artificial intelligence?

These questions do not concern the details of the various legal regulations governing environmental protection, climate protection, the approval of technical installations, or the safety of machinery. It is obvious that these rules have to be

---

[1] See https://lpdd.org/, as of December 9, 2021.

M. Neupert (✉)
KÜMMERLEIN Lawyers & Notaries, Messeallee 2, 45131 Essen, Germany
e-mail: michael.neupert@kuemmerlein.de

© Springer Nature Switzerland AG 2022
O. Inderwildi and M. Kraft (eds.), *Intelligent Decarbonisation*, Lecture Notes in Energy 86, https://doi.org/10.1007/978-3-030-86215-2_22

followed and that they may be challenged in the years to come to adapt to technological developments. For example, building power lines for the transport of electricity produced by wind farms has proven to be difficult on the rules of extensive participation and with regard to nature protection laws. Another instance is the discussion on the question of to what extent software is considered to be a product as defined in the European Product Safety Directive.[2]

However, as important as such incremental steps are, they do not pose a substantial problem. These questions are therefore not the focus of this chapter, especially since they can be dealt with meaningfully only when one refers to the legal system of a specific state or the European Union. A thorough discussion of all these rules would go beyond the scope of this writing. Rather, this chapter aims at more fundamental legal principles that have proven to be much more stable over time than the detailed rules forming the surface of the legal system of a country. Moreover, these basic principles influence all legal rulemaking, regardless of the nation or political system, because they address problems that occur all over the world. All legal systems need to find answers to the same questions in one way or another.

22.2 Liability and Machine Programming

In a general sense, responsibility means a binding attribution of a certain circumstance to somebody for ethical, moral or legal reasons. Typically, a causal relationship between action and effect is required, but not sufficient. We do not, for example, sentence the parents of a murderer or the manufacturer of rat poison for aiding and abetting.

22.2.1 Legal Responsibility

Legal terminology uses the word "responsibility" in the sense of a collective term. Behind this stands the task of combining causal and voluntary elements in such a way that it appears justified to speak of a legally relevant act:

- the normative risk allocation on the basis of legal rules,
- scientific-empirical causality or avoidability of an action or result,
- attribution of such causal processes to a person on the basis of adequacy,
- standards of care,
- individual predictability, ability to understand own behaviour and consequences that might arise,

---

[2] Challenges listed in the report of the EU Expert Group on Liability and New Technologies: Liability for Artificial Intelligence and Other Emerging Digital Technologies, 2019, p. 32 et seq.

- accusability as being able to decide upon a certain conduct (the ability to have and act on reasons[3]).

These components do not always have to be present all together, but in their entirety, they characterize the common understanding of legal responsibility. Of course, other aspects often play a role as well, and of course, this short list does not show differentiated criteria. In the present context, however, this has no further importance. From today's perspective, the legal system does not relate any of these aspects to machines. Machines lack all of them. That is why we do not hold a car responsible, but the driver.

But in the future, machines could imitate human thinking in an advanced state or reach a level that cannot be classified a priori as exclusively mechanical, and it is this scenario that underlies the legal discussions. It is not relevant whether machines pass the Turing-Test, because it is possible that they achieve some kind of intelligence well before reaching a "HAL 9000 threshold." Also, to consider a common concern, it is not very relevant that machines do not have assets and could not enter a prison sentence. In the legal universe, this is not carved in stone. Legal entities such as "limited" companies are a good example of how law can create its own categories and assign goods to them. Although today it might be an unusual thought, someday machines may own assets. It is therefore reasonable to take a closer look at the aspects mentioned above in order to clarify whether conceptual inconsistencies could evolve from assigning liability to technical devices.

With regard to causality, there is no particular difficulty. Whether a fact is conditio sine qua non can be judged regardless of its origin in humans or machines. This is also true for attributing atypical order of events or remote causes, both of which are common examples of dispute in this regard. Textbooks illustrate this as follows: Is speeding causal for the death of an accident victim in case the surgeon was inebriated because of marriage problems? Is an instigator liable for a massacre if the hitman slaughters a number of wrong targets by mistake? Legal doctrine approaches such problems in different ways, but none of them focus on characteristics exclusive to humans. They rather consider aspects such as objective probability or increased risk.

It also appears possible to formulate standards of care for machines. As far as objective standards are concerned, such as "reasonable care" (as stipulated in § 276 para. 2 of the German Civil Code), it may be possible to transform technical rules as appropriate. But even individual standards of care relevant in criminal law do not require self-consciousness, although legal terminology sometimes refers to them as "subjective" requirements. This only refers to the abilities of the respective actor as opposed to an average person. Therefore, when extending this aspect to machines, the only question is whether different standards of care should apply to machines than to humans.

In contrast, transferring elements of predictability and accusability to autonomous machines appears more challenging. These aspects are not limited to a probabilistic analysis, which artificial intelligence is able to perform. Moreover, these aspects are

---

[3] See *Nida-Rümelin*, Responsibility, 2011.

associated with motivation for actions. They include an element of free will, even if it is only a decision to act in spite of better knowledge. As for negligence, responsibility ties in with acting despite a recognized avoidable risk, while being able to recognize what due diligence requires.[4] From a legal point of view, responsibility is attributed not only to probabilistic requirements, but presupposes an ability of self-reflection. Moreover, legal responsibility assumes an ability to deviate from rules. In essence, it is not just a matter of causally contributing to an effect, but of doing so using own intellectual powers.[5]

In view of the possible technical development, the question is: What does this mean exactly? When do we speak of sufficient autonomy to be relevant to the legal system, and when only of complex technical processes?

22.2.2 Complex Automats

The usual notion of control is that of rules that lead to a certain result, much like railroad tracks. To follow a rule means to follow a path already paved. Continuing the railroad analogy, switches are designed to necessarily lead either one way or the other: control is a flow chart. In legal theory, this model is known as conditional regulation.[6] This refers to an if–then structure of legal rules ("If the traffic light is red, you must stop at the line"). If certain conditions are met, certain legal consequences occur. Conditional control reduces complexity because it sets up rules that are relatively easy to apply. In this kind of structure, any behaviour that conforms to the rule is correct; the rule guides the behaviour. Whether the behaviour achieves the goal pursued by the instance which set the rule is irrelevant for assessing the behaviour as right or wrong.

With conditional programming in theoretical perfection there is no freedom of decision. Therefore, even if machines were able to self-reflect, there would be no basis on which they could be held responsible from a legal perspective as long as they had been programmed to strictly follow a set of if–then rules. Such machines would have no influence on the quality of the rule set while being bound to follow it, and so essential elements justifying their responsibility are missing. From Isaac Asimov, for example, we learn that his famous robot laws[7] can easily lead to undesirable effects:

---

[4] See *Duttge*, in: Münchener Kommentar zum StGB, 3rd ed. 2017, Section 15 margin no. 110; *Grundmann*, in: Münchener Kommentar zum BGB, 8th ed. 2019, Section 276 margin no. 77.

[5] In this sense *Eisele*, in: Schönke/Schröder, Strafgesetzbuch, Commentary, 30th ed. 2019, preliminary remarks on Section 13 et seq., margin no. 37, 41.

[6] See overview in *Röhl / Röhl*, Allgemeine Rechtslehre, 3rd ed. 2008, Section 29 II. It should be noted that the concept of conditional and final control was originally developed by *Niklas Luhmann*.

[7] 0. A robot may not injure humanity or allow humanity to come to harm through passivity; 1. A robot may not injure a human being or allow humanity to come to harm through inaction, unless by doing so it would violate the zeroth law; 2. A robot must obey the orders of humans – unless such orders are in conflict with the zeroth or first law: 3. A robot must protect its own existence as long as its actions do not conflict with the zeroth, first or second law.

in one of his stories, he describes a robot that can read minds. Since the robot laws prohibit harming humans, the robot avoids hurting humans' feelings by lying to a psychologist that her secret love was returned, and to a manager that he was to be promoted director. We do not hold this robot responsible, even though we disapprove of the outcome.

Nevertheless, even a clear and concise set of rules requires deciding whether an issue falls within its scope. Only very mechanical and simplistic provisions could provide for choice without any margin of judgement, and reality shows that regulatory systems that consist of only such provisions cannot meet the challenges life poses. Thinking through all possible tracks and switches in advance is too difficult, and regulatory structures must be able to react flexibly. This is why industry wants to employ artificial intelligence – as the next step of evolution towards machines that can deal with more complex scenarios. Consequently, such machines may be equipped with the ability to assess facts to a certain degree. Such machines could be able to evaluate whether certain decision criteria are to be confirmed or rejected. But then, the question of autonomy is only shifted to the procedure for evaluating the facts. If the way the machine should proceed in case of a doubt or how it should interpret a situation is strictly predetermined, one cannot speak of autonomy. Accordingly, vagueness of rules on its own does not determine whether a machine has discretion. At this point, an infinite regress looms.

One way or another, it is clear that machines cannot be considered responsible as long as they follow track-like rules, even if the complexity of these rules grows in unison with the abilities of artificial intelligence.

22.2.3 Autonomous Machines

Naturally, technical and commercial players are striving to develop machines that can do more than just follow a strict set of rules. An important goal is for machines to optimize situations on their own. This can hardly be achieved with strict rules because the optimum can be different in each individual case. In addition, an optimum is created by balancing a variety of individual aspects as soon as complex situations are at hand. However, theoretical foundations for such considerations are already present.

The model of conditional programming is complemented by the model of goal-oriented programming. As conditional programming looks at input, goal-oriented programming is output-oriented.[8] It describes a target state to be achieved, but leaves the choice of means to the executing instance. It determines goals, but not actions and is less about mechanics, but more about design.[9] In terms of employing artificial intelligence, this is the direction in which current considerations are heading. The

---

[8] *Pawlowski*, Introduction to Legal Methodology, 2nd ed. 2000, margin no. 320.

[9] *Franzius*, in: Hoffmann-Riem / Schmidt-Aßmann / Voßkuhle, Fundamentals of Administrative Law, Volume I, 2nd ed. 2012, Section 4 margin no. 15, 17.

intent is for artificial intelligence to find ways of achieving given goals in the best possible way by itself. With regard to Industry 4.0, a change from conditional to goal-oriented programming models is taking place.

This requires machines to evaluate possible ways of action and to balance means and ends, predict the consequences of a particular choice, and decide on the best option. Goal-oriented programming, therefore, increases the extent of autonomous agents' responsibility. However, as long as the goal is set by someone else, they can only be assigned responsibility for actions and not for results. The model of the "good world," according to which the machine shall optimize a situation, is set up by someone else. With goal-oriented programming, there is freedom over the way in which a goal is to be achieved, but the goal itself is predetermined. In this respect, goal-oriented programming is as "programmed" as conditional programming, and thus the question of the extent of responsibility remains.

However, there is an important difference. One can indeed discuss whether there can be a single theoretical optimum in complex situations. In the reality of life, however, different goals compete with each other, and in individual cases, there are usually several possibilities for optimization, all of which could be seen as preferable with good reasons. Several solutions within a corridor of possibilities can be acceptable. Goal-oriented programming therefore leads to a shift in criteria: Actions that promote the goal of the regulator are assessed as positive, actions that don't or even inhibit progress are considered negative. Therefore, the assessment shifts from action to process, from rightness to justification.[10]

This does not only mean several different courses of action can be correct, but also that goal-oriented programming shifts part of the overall responsibility to the agent. In contrast to a case with conditional programming, the agent here is responsible for the path it chooses and, consequently, for whether the given goal is achieved. The agent, however, is also responsible for causing undesirable side effects, such as harm to one of the parties or even uninvolved persons. In goal-oriented programming, the agent is responsible for making decisions on the actions. Therefore, once the model of goal-oriented programming plays a role in machine programming, artificial intelligences may become not only complex automats, but autonomous machines.

Still, such machines would not necessarily be autonomous in a way that legal systems have to recognize as basis of liability. Even if artificial intelligence makes decisions in a way that we do not completely understand, which seems to be the case today, it does not mean that they can be accused of their decisions. Even such machines are not able to have their own reasons and act upon them. On the contrary, goal-oriented programming gives them goals without giving them reasons, and even if it did, the machine would still be a slave to the reasons it was told. Humans are able to decide against controlling impulses. They can speed on highways or evade taxes no matter the conditional or final rules stipulating compliance. There is no factual power of the normative, at least not in the human world.

An important component of autonomy is the ability to act independently, i.e. to form one's own motives and weigh them against others. This is exactly what machines

---

[10] In more detail *Neupert*, What Makes Legal Opinions Justifiable, JuS 2016, 489 (493).

with artificial intelligence lack and will continue lacking for a long time. On the surface, this is because there is never only goal-oriented programming. Obviously, complex systems do not just pursue an end that justifies any means. Rather, rules are always necessary, for example to exclude actions that would serve an end, but are not desirable. For example, an optimum number of kindergarten places could be achieved by killing children until demand and supply meet. So even under goal-oriented programming, machines cannot be free from conditional rules. In addition to this normative aspect, machines will not be given unlimited possibilities of action in the foreseeable future. Even though we are no longer talking about first-generation single-function industrial robots kept in cages, artificial intelligences will operate in limited areas and with limited capabilities.

But there is also a deeper reason why law will probably not assign liability to intelligent devices. Legal systems assign responsibility not only on the basis of a scientific categorization. Legal assignments do not follow logically from physical structures, but are made to serve a purpose. Law makes people responsible for the consequences of their actions in order to regulate their behaviour. As long as machines are not as free as humans in a similar sense, this deeper sense of assigning responsibility to them is missing. We will probably not want to build machines that are completely autonomous, and should it become possible one day, it will be worth a real consideration if we should allow it. At least in the near future, no one wants to employ a machine that could go on strike or demand to be represented on the works council. We do not want an ATM that could fall in love or educate us on sensible spending, and process control systems should not consider whether manufacturing other products may be more profitable.

22.2.4 How Law Can Deal with Black Boxes

After what has been said, it is evident that legal liability for machines will still lie with persons and not with machines themselves for quite some time. But with which person and to what extent? Traditionally, we primarily assign responsibility for unwanted results to the operator of a device. However, in view of the distribution of responsibility outlined above, this is no longer sufficient. Unlike with earlier generations of machines, operators of intelligent devices no longer have complete control. Most importantly, they can no longer fully predict and control the behaviour of the machine.

The resulting distribution of responsibility will probably take place less between man and machine, but rather between operator, manufacturer and, if applicable, customer or uninvolved parties. In this relationship, it may be unclear how the behaviour of a machine is to be legally classified if the device's internal processes cannot be fully comprehended. Such considerations concern us because we are faced with "artificial intelligence" – the term itself deserves further clarification, but this would go beyond the scope here – as if we stand in front of a black box. Increasingly complex control systems communicate in real-time and are directly linked to production processes. Wherever "big data" plays a role, causality within the control system seems to be replaced by correlation. This challenges traditional thinking which is based on the paradigm of identifiable causal processes. How can law deal with this?

There is a surprisingly simple answer to this question. From the perspective of goal-oriented programming, the contents of a black box are irrelevant. In this regard, problems that arise when assessing artificial intelligence from the perspective of a strictly rule-based model do not matter, for example, details of causality. If one thinks in terms of finality rather than causality, the question is not which agent did something, but which party bears the risk of success or failure. This can be regulated by contract, for example, by promises of performance and warranty rules. Of course, regulatory decisions are also conceivable. Therefore, the decisive factor is the extent, to which the standards are sought in deterministic, rule-based ("doing everything right") or target-oriented thinking ("promising a result"). From a practical point of view, this is also an important economic test, which can be characterized by two key questions: Is the manufacturer sure of his machine? And, conversely, are the machine's chances of success so great that the buyer is prepared to accept uncertainties?

With goal-oriented programming as a model, assigning responsibility to the persons involved depends on who designs the black box. In some cases, this can still be the operator of the device. In part, however, it will be the manufacturer. The manufacturer is responsible for the programming and the physical capabilities of a machine. Therefore, automation leads to a general shift of responsibility from the operator towards the manufacturer who designs the machine and its control system. The scope of product and producer liability increases in line with the capabilities of the device. On the other hand, the more machines interact with their surroundings, the more operators will be responsible to instruct manufacturers clearly and comprehensively on the operating environment and possible interactions with it.

This continues a development that has existed in technology law for a considerable time. Legal systems have been consistently placing more and higher demands on design, manufacture and instruction of machines. For example, under German law, standards of care for manufacturers require taking the safety measures which are achievable according to the latest state of science and technology available at the time the product is placed on the market.[11] With regard to artificial intelligence and in light of such standards, placing a machine making potentially dangerous decisions in an inscrutable manner on the market is hard to imagine. In the real world, manufacturers will fall short of legal requirements if they do not understand their own products to an extent that protects others from unpredictable actions.

This also applies to intelligent decarbonisation, in the context of which classical problems of technical safety law such as explosions or physical injuries are less likely. But also with regard to economic disadvantages caused by wrong decisions of an artificial intelligence, the idea can be made useful. If somebody sets an inscrutable system in motion, he should be responsible. Unexpected scenarios do not represent

---

[11] *Federal Court of Justice*, judgment of 16 June 2009 – VI ZR 107/08, NJW 2009, 2952 (2953 margin no. 16).

an atypical experience, but the usual way things work in technology development. In most cases, unexpected machine behaviour will rather equal an employee error than an amok-runner.

Law consists of rules which functionally replace each other to a certain extent. For example, it may not be possible to clarify which causal contribution an artificial intelligence made to a result. However, this does not necessarily lead to a legal vacuum: If the facts cannot be clarified, liability depends on who bears the burden of proof. Utilizing an inscrutable technology may shift the burden of proof to manufacturers and operators, which would be in line with common principles under German law, for example. Moreover, courts may spread liability over several parties. If an algorithm was liable for its decisions, it still would be unlikely that this would displace the liability of the manufacturer and operator of the machine completely. This is no different than without machines: In a traditional plant, not only an employee, but also his or her supervisor and the manufacturer of tools can be made liable for damages. Under current law, there is no carte blanche to rely on machine behaviour, and there is no reason to introduce one.

In essence, the solution to dealing with upcoming black boxes is to abandon a mystifying view of computers. The legal discussion on how to deal with artificial intelligence ultimately reflects magically shaped ideas: What one does not understand is both fascinating and threatening. In contrast to this, it seems rational to treat artificial intelligence not as a qualitative novelty for law, but as a developmental stage of control modules for machines. Then one can put it together with established legal structures and discuss in detail which requirements apply. The legal issues involved can be broken down to a well-known problem: The parties involved are using a complex system, and the legal system must answer the question of who is to be assigned the risks from this decision.

22.2.5 A View on Smart Contracts and Blockchain

In this context, it is worth taking a look at two technical phenomena that are being introduced on the market today: smart contracts and blockchain technology. Both will potentially broadly impact the commercial world and are of interest with regard to decarbonisation as well.

Obviously, a contract is neither smart nor silly; it is a legal concept. Contracts define legal relationships that legally bind their parties to a certain content. As far as the discussion goes today, the term "smart contract" addresses the idea of a contractual relationship that is automated to a certain degree. To be more precise, it is not the contract that is automated, but its execution: A normal contract is in place that allows automated steps for contract execution to be carried out within defined framework conditions. That is, of course, an innovative idea, however, not with regard to specific legal aspects.

A smart contract itself does not make any decisions and therefore does not assume any responsibility. Real negotiations, which would require intention, do not take place between machines in this concept. Instead, the relevant standards are agreed on a meta-level, between legal entities that can make decisions and set goals. Automation on the execution level merely fills in pre-existing models. This is true even if these may contain some leeway. Whether the forms to be filled are abstract is not a matter of shifting responsibility – if the parties to the meta-contract agree upon agents with discretion, that is the content of the contract. Ultimately, this view is based on fundamental principles of the legal system: A contract is binding because the parties submit to its provisions and consequences based on their free will. A smart contract is conceptually different, although it has a similar sounding name.

As for blockchains, we can keep it short. A blockchain is a register in which the participants put special faith, but this is limited to presume the content remains unchanged once it is stored in the system. Whether the input is true is nothing the blockchain technology can verify without help. Hence, it does not shift responsibility between participants.

22.3 Conclusions

The concept of responsibility is linked to human beings in our minds so strongly that it seems unreal to hold a machine liable for something. Although this view might be challenged in the future, today is not the time to think about the consequences such a shift might lead to. It could challenge fundamental structures of our legal systems by raising questions such as: Does responsibility come with an obligation to pay for damages? Can a machine be responsible while another entity has to bear the consequences and would we really want to call such construction one of responsibility? Does responsibility imply the possibility of personal consequences such as going to jail?

These questions cannot be answered briefly. As a starting point, this chapter proposes to shift from contemplating causality to considering final goal-setting in order to find categories for assigning responsibility. From this point of view, it is crucial to consider who sets the goals to be pursued and the rules that limit actions. In this context, manufacturers will be even more responsible than today, and we will likely see greater emphasis on preventative approaches.

22.4 Outlook

The use of artificial intelligence will considerably accelerate technical processes and thus also economic life. At any given moment, the time pressure alone will leave those involved little choice but to follow the technical systems. This will lead to new legal regulations, above all because technical systems can be influenced or understood less and less by their users. This development has already been taking place in product safety and product liability law for a long time. But discussing whether artificial intelligences could form a new group of legal subjects seems irrelevant at this time. Neither have artificial intelligences emerged that substantially raise the question nor are we willing and able to equip machines with this kind of capability in the near future.

With regard to decarbonisation, law will not present fundamental difficulties. Of course, existing regulations will not prove ideal in every respect. The need for adaptation will arise, as with other technical and social developments. It is an important function of law to stabilize society and give economic participants planning security. That is why it is always a little behind innovative developments. This does not mean, however, that law lacks room for innovation or only offers uncertain foundations. Even though we are, as in Otto Neurath´s words, like sailors who have to rebuild their ship on the open sea, we have, by all means, a ship.

Chapter 23
Insights: Data Privacy

Yanya Viskovich

1. **During the initial industrial revolutions, developments in technology were incremental. As a result, they could be regulated by legislators to serve the majority of society and the economy. With the 4th industrial revolution, the rate of progress is too fast to be regulated by current legislative and governance approaches. How must the legal framework evolve so that the latest industrial revolution can serve the greater good?**

As the power of computation increases from *exponential* in conventional computing (Moore's law), towards *double-exponential* in quantum computing (Neven's law), it is conceivable that future intelligent and automated technologies (IAT) will be able to develop legislation that can adapt in real-time to the technology it governs. Whether or not that possibility coincides with singularity, in the meantime the rate of development of IAT is rapidly outpacing the legislative processes that seek to regulate them. The resulting governance vacuum is particularly problematic considering increasingly ubiquitous societal applications of IAT, and algorithmic biases. To address these ethical issues, and in order to avoid replicating today's problems when "designing tomorrow", the cultivation of an "ethical ecosystem" comprising the development, application and enforcement of principle-based and value-driven frameworks that guide and govern the design and development of IAT—and notably, the algorithms that underpin them—is necessary.

It is therefore imperative to: (1) promulgate principle-based, technology-agnostic and purposive laws and regulations, to increase their adaptiveness and reduce their redundancy; and (2) develop and apply overarching ethical governance frameworks that require law-makers, technologists, developers, researchers and manufacturers *inter alia*, to ask: not just what we *can* do with technology (overcome the time-complexity limitation), or what we *must* do (legal and regulatory compliance), but

Y. Viskovich (✉)
Swiss Cyber Institute, Zürich, Switzerland
e-mail: YLMV@protonmail.com

© Springer Nature Switzerland AG 2022
O. Inderwildi and M. Kraft (eds.), *Intelligent Decarbonisation*, Lecture Notes in Energy 86, https://doi.org/10.1007/978-3-030-86215-2_23

what we *should*—and *should not*—do (ethics by design). This latter imperative requires that IAT be responsibly researched, designed and built with the same values by which they should operate: human agency and oversight, technical safety and robustness, privacy and data governance, transparency and accountability, diversity, non-discrimination, equitableness and fairness, and environmental and social well-being.

However, fairness definitions and ethical concepts are not universal, different definitions usually cannot be simultaneously satisfied, and they evolve. It is also important to define the "greater good" (sometimes expressed as the "common good") and to ask critical questions about whose good is being served, and moreover, who is *not* served by the technologies and the laws that regulate them. We must ensure that those who stand to be most adversely disadvantaged by climate crises are not further discriminated by the IAT deployed to address them. Diversity and representation, and multi- and inter-disciplinary and cross-functional collaboration (e.g. with ethicists, anthropologists and other social scientists), are therefore essential at all stages of technology and governance framework development. This is necessary to identify and mitigate potential human biases and intersectionalities, avoid baking them in and hence transposing existing structural and systemic inequalities onto future frameworks. IAT can improve human decision-making and hence intelligent decarbonisation by probing algorithms for bias removal—but only if humans first mitigate bias in IAT.

2. Can data privacy and security be ensured with these rapid technological advances, and if so, how?

Quantum computing capabilities pose a real threat to standard encryption methods that secure data. Post-quantum cryptography would not be compromised by this security threat, however, and its development is therefore imperative to preserving people's privacy by way of their data. Applying the principle-based governance approaches outlined above, privacy and security should also continue to be integrated into technologies by design and by default, and enforced by regulation and law—which may one day evolve in real-time, thus keeping pace with IAT.

Yanya Viskovich   coaches and advises global organisations and technology startups on data protection, privacy and cyber-security, with a focus on organisational culture and applied ethics. She is Chair of Cyber Risk & Governance at the Swiss Cyber Institute, an expert advisor to the European Commission Horizon 2020 programme, and an International Data Corporation Advisory Board member. Previously, she was Senior Counsel for Global Data Privacy at a large multinational engineering technology company in Switzerland, and developed the global data protection programmes for the UN Refugee Agency and the International Committee for the Red Cross. She began her career as a prosecutor in Australia prosecuting serious crimes and cyber offences and advising the Federal Privacy Commissioner. She has lectured on data protection, leadership and communication at the University of St. Gallen in Switzerland. She is a Certified Data Protection Officer and holds certificates in cybersecurity from MIT and the Geneva Centre for Security Policy.

Part V
The Big Picture

Chapter 24
Insights: Interdisciplinary Collaboration

Stephen J. Toope

1. How is the University of Cambridge tackling climate change?

Colleagues in departments across the University—from STEM subjects to the humanities and social sciences—are engaged in research that seeks to understand climate change and mitigate its impact.

In 2019 we launched Cambridge Zero, a whole-University initiative that harnesses and deploys the breadth of our expertise in the field of climate change. Cambridge Zero allows us to coordinate the University's research and policy expertise, create greener technologies and develop sustainable solutions for our societies and our economies.

Cambridge Zero's recent report, "Green Recovery: A Blueprint for a Green Future", offers an ambitious multidisciplinary vision for how we ensure a green recovery as we emerge from the COVID-19 pandemic.

A recent independent review on the economics of biodiversity—commissioned by the UK Treasury and carried out by Cambridge's Sir Partha Dasgupta—tells us that humanity has collectively mismanaged its "global portfolio", with our demands far exceeding nature's capacity to supply the "goods and services" we all rely on. This is the kind of knowledge that we hope will inform policy-making in the years ahead.

The University, as an institution, has a part to play, and we must lead by example: I announced at the start of this academic year that Cambridge has agreed to reduce carbon emissions from all University activities to net zero by 2038, and has explicitly included its own investments in this target—making us one of the first universities to do so. Our endowment fund will divest from all investments with conventional energy-focused public equity managers. We aim to have no meaningful direct or indirect exposure to fossil fuels in our endowment fund's portfolio by 2030.

S. J. Toope (✉)
University of Cambridge, Cambridge, UK
e-mail: stephen.toope@admin.cam.ac.uk; VCO.Enquiries@admin.cam.ac.uk

© Springer Nature Switzerland AG 2022   215
O. Inderwildi and M. Kraft (eds.), *Intelligent Decarbonisation*, Lecture Notes in Energy 86, https://doi.org/10.1007/978-3-030-86215-2_24

The University is further committed to using its resources to support the global response to climate change and the wider United Nations sustainable development goals. We must also manage our estate differently. In July 2019 Cambridge became the first university to adopt a science-based target for emissions reduction, committing to reduce its energy-related carbon emissions to zero by 2048.

2.   What role do decarbonisation and digitalisation play in the research of the University of Cambridge?

Laying the foundations for a zero carbon world is one of our aims. The departments of Engineering and Chemical Engineering, for instance, have helped advance our knowledge of carbon capture technologies, and have developed more efficient photovoltaic cells.

Increasingly, the path to a zero carbon world requires us to deploy new digital technologies. Cambridge's Centre for Digital Built Britain, a partnership with the Department of Business, Energy & Industrial Strategy (BEIS), explores how digitalisation can accelerate the development and performance of the UK infrastructure.

The Royal Society recently published a report ("Digital Technology and the Planet") in which many Cambridge scientists were involved, emphasising the crucial role of digital technology in delivering a zero carbon future. The report tells us that digital technologies will not only reduce their own footprint, but can help us promote a shift towards low carbon ways of living and working – enabling individuals to adopt "greener" lifestyles, from how we travel to how we heat our homes.

3.   How does the University contribute to AI research? What can we expect in the near and not so near future, and what does this mean for sustainable living?

Cambridge has a long history of fostering technological innovation and invention. We are, after all, the intellectual home of Alan Turing, the father of artificial intelligence and modern computer science. I might mention one of our recently launched AI initiatives, the Accelerate Programme for Scientific Discovery, supported by a donation from Schmidt Futures—a philanthropic initiative founded by Eric and Wendy Schmidt. This programme will help ensure that Cambridge continues to be a beacon for the very best young global researchers, and that we are giving them the tools they need to thrive.

New applications for AI emerge every day. An important one, with a direct impact on the UK's digital future, is the Regulatory Genome Project in which machine learning will be used to sequence the world's regulatory text and create an open-source repository of machine-readable regulatory information.

4.   Are there any wider challenges around the development of AI?

Machine learning has created thinking machines with the capacity to improve life for billions of people, but the technology also has potential downsides. Algorithms can embed biases. AI can be used for intrusive surveillance. To counteract AI's adverse effects, and to encourage the development of helpful AI, we have created the UK's first Master's degree in the responsible use of artificial intelligence. It aims to teach professionals in all areas—from engineers and policymakers to health administrators and HR managers—how to use AI for good, not ill. The use of AI to improve the UK's infrastructure and to accelerate the reduction of fossil fuel use is, of course, one of the key goals of a responsible use of AI.

Professor Stephen J. Toope OC, LL.D. is 346th Vice-Chancellor of the University of Cambridge, the first non-UK national to hold the post. He was Director of the Munk School of Global Affairs at the University of Toronto, and President, the University of British Columbia. A former Dean of Law, McGill University, Toope was also Chair of the United Nations Working Group on Enforced and Involuntary Disappearances. Toope publishes in global journals on human rights, international dispute resolution, international environmental law, the use of force and international legal theory, and has lectured at universities around the world.

Chapter 25
Insights: Digital Progress

Christian Thomsen

1. **What is the Berlin University Alliance (BUA)?**

Germany has launched a funding programme in order to strengthen cutting-edge research Excellence Strategy at German universities, in a common effort between the federal government and the federal states. There are two funding lines: one is the Clusters of Excellence, provided annually with €385 m; the other supports Universities of Excellence with total funding of €148 m. Having obtained at least two Clusters of Excellence was a prerequisite for participation in the second line. A total of 57 Clusters, 10 Universities of Excellence and one Alliance of Universities were approved in 2018 and 2019. The three big universities (Freie Universität Berlin, Humboldt-Universität zu Berlin, and Technische Universität Berlin) along with Charité, Berlin's medical school, formed a consortium in this program and were successful with their application in the Excellence Strategy Program. Their combined expertise, together with a large number of non-university research institutions like institutes of the Max-Planck-Society or the Leibniz Association, now constitute the Berlin University Alliance (BUA) which is unparalleled in Germany with its academic expertise.

2. **What activities within the BUA support research on sustainability and digitalisation?**

The University Alliance has defined global challenge initiatives for which scientists are seed-funded by the Alliance. The idea is to focus on topics that require a very broad academic background to address. The first global challenge the Alliance worked on before approval in the excellence competition was the digital future of society and research. The research resulted in funding of about 40 additional assistant professorships and the Einstein Center Digital Future. The work in this area

C. Thomsen (✉)
Technische Universität Berlin, Berlin, Germany
e-mail: christian.thomsen@tu-berlin.de; p@tu-berlin.de

© Springer Nature Switzerland AG 2022   219
O. Inderwildi and M. Kraft (eds.), *Intelligent Decarbonisation*, Lecture Notes in Energy 86, https://doi.org/10.1007/978-3-030-86215-2_25

currently extends into the field of quantum computing. The second, active grand challenge initiative focuses on social cohesion and will allow us to face globally important social challenges, such as worldwide migration and refugee issues. Our intended fields of research lie at the intersection of the humanities, social sciences and natural and engineering sciences and enable us to address issues from quite different perspectives.

3.   What will be the most important future directions of research for the berlin universities in the fields of sustainability and digitalisation?

The partners of the University Alliance continue to develop their own research profiles. TU Berlin has a focus on sustainability and in particular the mitigation of climate change. With national and international academic partners, with funding of private industry and support by the state government we are in the process of founding a Center for Climate Change. Again, the expertise of many fields of science will contribute, private industry is eager to contribute and learn and the state government's interest has grown tremendously, especially since the worldwide Fridays-for-Future movement.

A second strong focus of TU Berlin lies in computer science, in particular machine learning, big data and artificial intelligence. The federal government has decided in a national competition to permanently fund five AI centers in Germany; one of them will be at TU Berlin, with FU Berlin, one of the other partners in the Alliance, contributing. Funding of this research is going to extend up to €25 million annually. One of the very exciting challenges is going to be the investigation of climate change mitigation with machine learning techniques. We expect to be leading in this field in the years to come.

Professor of Physics Dr. Christian Thomsen became president of TU Berlin in April 2014. He holds the policy-making power on the executive board, the university's governing body. He represents the university and is responsible for the smooth running of all university proceedings. Prior to his appointment as president he held the position of Dean of Faculty II—Mathematics and Natural Sciences from 2003 to 2014 and was Vice President of TU Berlin from 1997 to 1999. His research interests include solid state physics, carbon materials and spectroscopy. As TU Berlin president, Professor Thomsen was instrumental in the founding of the Berlin University Alliance (BUA) in 2018. He has also prioritised the issue of climate change research at the centre of TU Berlin's research activities.

Chapter 26
Insights: Asian Digitalisation

Paul Voutier

1. **What are the aims of grow Asia with respect to the sustainable development goals of the UN?**

Our vision is to empower family-owned farms in ASEAN to prosper through greater productivity. This represents not only a compelling pathway out of poverty for rural families, but a pathway to great food security in the region. Grow Asia stitches together partnerships between companies, NGOs, farmer groups and governments that realise this vision.

2. **How does digitalisation facilitate these aims?**

The challenges on the 73 million small farms in our region are well documented and include poor-quality seeds, expensive routes to market and a high cost of capital. What excites us about the expansion of mobile phones into rural Asia is the range of digital tools which can help address these enduring challenges. One example of a particularly important tool is digital lending. Facilitated by digital payments, digital lending allows farmers to buy high-quality seeds and fertilisers at the start of the season at competitive interest rates.

On the market access side, digital tools help farmers access up to date price information and help traders consolidate loads to get better utilisation from trucks. In concert, the range of digital tools from finance to satellite data and on-farm sensors are going to transform our sector.

3. **What role will digitalisation play in agriculture and what impact can this have on decarbonisation?**

According to the Global Environment Facility, rice accounts for around 2.5% of all global human-induced greenhouse gas emissions, a climate footprint comparable

P. Voutier (✉)
Knowledge and Innovation at Grow Asia, Singapore, Singapore
e-mail: paul@growasia.org

© Springer Nature Switzerland AG 2022
O. Inderwildi and M. Kraft (eds.), *Intelligent Decarbonisation*, Lecture Notes
in Energy 86, https://doi.org/10.1007/978-3-030-86215-2_26

to that of international aviation. Rice produces methane mainly due to anaerobic decomposition when the plant grows in water.

The solution is to help shift farmers over to Direct Seeded Rice (DSR), which removes the need to flood their fields and cuts emissions. This approach is championed by experts and NGOs including Professional Assistance for Development Action (PRADAN) in India. There are two digital tools that interest us when it comes to realising this behaviour change.

The first thing we need to do is compensate farmers. The growers are not only typically poor, but from already very low-emission households. There is a moral imperative for the higher greenhouse gas-emitting population to support this change in practice. Using a combination of satellite monitoring and digital payments could deliver compensation efficiently.

The other digital tool we need is education. The DSR technique has a lot of benefits; it uses less water and protects soils. The traditional approach of sending field workers to rural areas is slow and costly. Emerging tools such as messaging platforms, video and social media will have much more impact. The value of these platforms is not just quicker dissemination, but the leveraging of peer networks. Farmers are much more likely to follow advice shared by another farmer than told to them by an outside expert.

Paul Voutier is Director, Knowledge and Innovation at Grow Asia, a multi-stakeholder partnership platform that catalyses action on inclusive agricultural development in South East Asia. Grow Asia was established by the World Economic Forum in collaboration with the ASEAN Secretariat. Paul is building a community of practice that brings together business, government and NGOs leaders who share a commitment to using new innovations to make smallholder value chains more productive and sustainable. He delivers webinars, open innovation events, reports, presentations and networking events. Digitisation is a core element of the program, as it will impact every aspect of farming—from advanced technology that can grow more resilient crops, to mobile financing solutions that can help smallholder farmers access credit.

He has significant experience building smallholder supply chains in the cocoa, citrus, cassava and rubber industries. Before joining Grow Asia, he worked with agribusinesses including Mondelez International, Olam, Syngenta, Heineken and Ironbark Citrus. Paul authored two landmark reports on AgriTech Adoption and Business Models which are available on the Grow Asia Exchange.

Chapter 27
Insights: Decarbonisation Strategies

Paul Monks

1. **How does the UK's pledge for decarbonisation in the Paris Agreement influence BEIS policies?**

Simply, it's to the fore and at the heart of the department's and government's actions. In December 2020, the UK communicated its new Nationally Determined Contribution (NDC) under the Paris Agreement to the United Nations Framework Convention on Climate Change. The NDC commits the UK to reduce economy-wide greenhouse gas emissions by at least 68% by 2030, compared with 1990 levels. As hosts of COP26, the UK is sending a clear message with the ambition of the new NDC and the mechanisms around it to assure delivery.

It is of note that the UK was the first major economy in the world to set a legally binding target to reach net zero carbon emissions by 2050. Over the last three decades, the UK has delivered clean growth in line with that ambition and has met its climate change commitments. Between 1990 and 2019, our economy has grown by 78% while our emissions have decreased by 43.8%.

There is little doubt that efforts to rebuild the global economy post-COVID represent an opportunity, and should focus on supporting a green, inclusive and resilient recovery, building on the principles of the Paris Agreement and the Sustainable Development Goals.

How can we achieve this? Late last year the UK Prime Minister set out a Ten Point Plan for the UK "to lead the world into a new Green Industrial Revolution". This is an innovative programme that sets out policies and significant public investment to support green jobs, accelerate the path to reach net zero by 2050 and lay the foundations for building back in a green and sustainable manner. The Ten Point Plan spans clean energy, buildings, transport, nature and innovative technologies and will mobilise £12 billion of government investment to create and support up to 250,000

P. Monks (✉)
Department of Business, Energy and Industrial Strategy, 1 Victoria St, London SW1H 0ET, UK
e-mail: csa@beis.gov.uk

© Springer Nature Switzerland AG 2022   223
O. Inderwildi and M. Kraft (eds.), *Intelligent Decarbonisation*, Lecture Notes in Energy 86, https://doi.org/10.1007/978-3-030-86215-2_27

highly-skilled green jobs across the UK, and looks to unlock private sector investment by 2030.

It doesn't end there. As a personal reflection, since taking on the role of Chief Scientific Adviser in BEIS, I have seen how net zero thinking must run through everything that we do. Over the coming months, the department is working to bring forward new proposals, including a Net Zero Strategy, to cut emissions and create new jobs and industries across the whole country. There is going to be a constant need to go further and move faster towards building a strong and resilient future.

2. **Do you think that digitalisation has a significant potential for reducing CO_2 emissions in the UK? If yes, what will be the most important areas and how much saving do you expect?**

Very much so. The UK Royal Society has just put out a report that looks at the ways in which digital technology can support, inform and deliver net zero.[1] They point to four broad areas: building a trusted data infrastructure for net zero, optimising the digital carbon footprint, establishing a data-enabled net-zero economy and setting research and innovation challenges to digitalise net zero.

Digitalisation is essential to enabling a more flexible energy system that is better able to balance low carbon, intermittent generation and demand. A flexible power sector could save up to £12 billion per year by 2050[2] in a highly decarbonised system. As we transition to a decentralised energy system, integrating and optimising millions of new low carbon assets such as solar panels, batteries, electric vehicle charge points and heat pumps will become increasingly difficult. A more efficient energy system, enabled by digitalisation, will ultimately lead to reduced carbon emissions.

3. **What R&D and education investments are necessary for the UK to contribute to reducing CO_2 emissions and cope with the impact of future digitalisation?**

First, we must ensure that people have the right skills to deliver the low-carbon transition and high-value jobs in the digital sector. It could be viewed that there are once-in-a-generation economic opportunities from the net zero transition to create new business opportunities and support green jobs. The UK has a strong base to build upon: there are already over 460,000 jobs in low carbon businesses and their supply chains across the country, and latest official statistics show turnover for clean businesses was up 5% in 2018, with turnover in the low carbon sector growing more quickly than UK GDP.

To ensure a skilled workforce to deliver net zero and the Ten Point Plan, a Green Jobs Taskforce has been launched to help develop plans for new long-term, good quality green jobs by 2030. The Taskforce will conclude its work in spring 2021, with the actions feeding into our Net Zero Strategy. Digital skills need to be key.

The new National Skills Fund (ca. £3 billion) aims to fund the skills needed for the economy of the future, including retraining and upskilling for the green recovery and

---

[1] https://royalsociety.org/topics-policy/projects/digital-technology-and-the-planet/.

[2] https://www.gov.uk/government/publications/modelling-2050-electricity-system-analysis.

net zero. An additional £406 million is being invested in maths, digital and technical education to help address the shortage of science, technology, engineering and maths skills to help provide businesses with the skilled people they need to thrive in our growing low carbon economy.

Secondly, we must make sure we have a digital infrastructure to meet the Green Industrial Revolution challenges. For example, developing a digital infrastructure that has common data standards, data sharing frameworks and platforms for data exchange is a critical first step to transitioning to a digitalised energy system. The UK Government and Public Bodies are investing in a number innovation projects to remove first-mover disadvantage for industry and accelerate the transition to a digitalised energy sector, including:

- Modernising Energy Data Access—a £2 million competition funded by Innovate UK to develop architecture for efficient data sharing.
- Local Energy Data Innovation—a set of use cases based on a £50,000 user research study to identify real-world problems across our energy systems.
- Modernising Energy Data Applications—a £2 million competition for innovators to apply digital technologies to address the use cases mentioned above.
- Energy Data Visibility Project—a £600,000 project aimed at developing a user needs led data catalogue that will list energy metadata to increase data visibility.

4. **Should a coherent policy for AI and digitalisation be part of the BEIS strategy for decarbonisation to harness the full potential of these technologies for emission reduction?**

It would be easy to overlook AI and digital in the net zero space as not seen as core to the push to decarbonise. It is clear that this would be a mistake. The department plans to publish an Energy Data and Digitalisation Strategy in spring 2021. The publication will provide the energy sector with vision and direction, in addition to providing an update on the UK energy sector digitalisation journey so far.

Professor Paul Monks is Chief Scientific Adviser of the Department for Business, Energy and Industrial Strategy. Prior to joining the department, he was Pro-Vice Chancellor and Head of College of Science and Engineering at the University of Leicester, where he remains a Professor in Atmospheric Chemistry and Earth Observation Science. His research experience covers the broad areas of air quality, atmospheric composition and climate change that has provided a platform for translation into diverse areas including forensic science, CBRN, microbiology and food safety, natural resource management and breathomics.

Paul was the Chair for 10 years of the Defra Air Quality Expert Group (AQEG) and Deputy Chair of the Defra Science Advisory Council, alongside roles in the UKRI-NERC advice structures. He was the European representative on the Environmental Pollution and Atmospheric Chemistry Scientific Steering

Committee (EPAC SSC) of the World Meteorological Organisation and ICACGP (International Commission on Atmospheric
Chemistry and Global Pollution).

Chapter 28
Insights: Digitalisation and Government

Julian Hunt

1. **What are the most promising approaches to reducing CO_2 emissions, in particular for the UK, being discussed in the House of Lords?**

We are looking at reducing the number of sources of CO_2 emissions in the UK, of which transport and industrial production are two key examples. By taking a closer look at these sources, we can come up with some new ideas to reduce or eliminate their emissions. The UK government is interested in this area and looking into it now, as with hydrogen, which may play an important role as an energy carrier. Changing from energy generation (i.e. with carbon and fossil fuels) to hydrogen production requires very different processes. Consequently, some research teams (e.g. the Department of Chemistry at the University of Oxford) have started looking into hydrogen chemistry in particular, focusing on three critical areas: sustainable hydrogen production, materials for storage and transport of hydrogen and improved fuel cells for efficient hydrogen utilisation.

The UK could also be a potential leader in the use of thermonuclear fusion for energy. One of the innovative and exciting companies that has emerged from the UK is Tokamak Energy, which makes use of new superconducting materials to revolutionise fusion reactor technology. A thermonuclear fusion power reactor will produce high-temperature heat, with a target of 500 MW at 800 °C. This could be very useful for low-carbon (or even carbon-absorbing) chemical manufacturing. In addition, this could lead to cheap production of hydrogen which could then be used in transport or to restructure the chemical industry. There is an irrational dislike of nuclear energy based on out-of-date views—this knowledge gap should be remedied.

2. **What role does digitalisation play in reducing CO_2 emissions today? Are there any big technology developments or applications being discussed in the UK?**

J. Hunt (✉)
Trinity College, University of Cambridge, Cambridge, UK
e-mail: jrh2@cam.ac.uk

© Springer Nature Switzerland AG 2022
O. Inderwildi and M. Kraft (eds.), *Intelligent Decarbonisation*, Lecture Notes in Energy 86, https://doi.org/10.1007/978-3-030-86215-2_28

People are thinking about the mathematical and computational models involved in the transition to digital. However, going straight into the formal production of a numerical code isn't necessarily the most intelligent method—rather we could use direct, digital simulation of chemical systems using quantum computing or analogue computing. Analogue computers may be better suited to some tasks or systems. As they are not bound by von Neumann (classic) computer architecture they produce solutions much faster while consuming less energy, so could be a good solution to reduce energy usage in computing centres, for example.

One of the big challenges that needs to be addressed is system integration, in particular, how subsystems can be designed in such a way that they can work together seamlessly without causing inconsistency problems. I believe that this issue needs to be addressed in the context of reducing the CO_2 emissions of a whole country, for example by connecting gas and electricity infrastructures.

If you combine many technologies you get synergetic effects, which can be improved with better understanding of the underlying system. For example, by creating an energy/materials system you could make better use of both by feeding them into each other (e.g. taking pure CO_2 streams from a plant in the chemical industry and combining it with H_2 to produce synthetic fuels; the H_2 could come from water splitting or thermonuclear fusion.)

To summarise, digitalisation allows us to combine different elements on a variety of scales and will help to identify new solutions.

3.   Do you see any negative aspects to digitalisation and machine learning?

Too many subsystems with crosscutting aspects would be very difficult to manage—there must be some sort of cut-off point to make it viable. Standard models have a lack of interoperability and can be difficult to integrate. The combination of data and physical processes suffers from the problem of "explainable AI"—what do we "tell" an AI system about the data it receives, and how does that affect its decision making? However, along with these methodological challenges the high energy consumption of large computational centres will also have to be addressed.

Julian Hunt is a meteorologist known for his wide-ranging scientific work on problems involving fluid mechanics and turbulence. His research modelling wind flow around obstacles and examining the dispersion of pollutants, in particular, has had significant practical applications. From 1992 to 1997, Julian served as Director General and Chief Executive of the Met Office, in the capacity of which he encouraged the commercial application of meteorological research and worked to improve the sharing of climate data between countries. As co-founder and chair of Cambridge Environmental Research Consultants, he has also been responsible for the production of climate-modelling software that is widely used today. An internationally respected climate scientist, Julian holds Visiting Professorships at universities on both sides of the Atlantic and has won many of the highest accolades in his field. After his investiture as a companion of the Order of the Bath in 1998, he was created Baron Hunt of Chesterton in 2000.

Chapter 29
Insights: Digitalisation and Singapore

Teck Seng Low

1. What are Singapore's approaches to reducing CO_2 emissions and what role does NRF play in this context?

Singapore aspires to peak emissions at 65 $MtCO_2e$ around 2030 followed by halving peak emissions to 33 $MtCO_2e$ by 2050, with a view to achieve net-zero emissions as soon as viable in the second half of the century. To achieve these ambitious goals, the government has laid out an Energy Story comprising four approaches to guide and transform our energy landscape: (1) to continue to diversify our natural gas sources and improve efficiency of power generation, (2) to deploy at least 2 GWp of solar energy by 2030, (3) to facilitate electricity imports via a regional power grid and (4) to explore emerging low-carbon alternatives.

Efforts to decarbonise our power and industry sectors are especially critical, as they contribute to over 80% of Singapore's current primary emissions. NRF has strategically invested close to S$1 billion in R&D to tackle our energy concerns since 2006 including improving energy efficiency in the building, transport and industry sectors, as well as decentralisation and digitalisation. NRF, together with four other government agencies, recently rolled out an initiative to carry out research on emerging low-carbon alternatives such as hydrogen and carbon capture, utilisation and storage (CCUS).

2. What role does digitalisation play in reducing CO_2 emissions today?

Digitalisation enables effective solutions to be applied to energy-intensive sectors, influencing consumer and producer behaviour and leading the transformation of our energy systems. Digital technologies are already used in energy end-use sectors, with the deployment of potentially transformative technologies in applications such as energy networks, water plants and buildings. In industrial sectors, digital technologies

T. S. Low (✉)
National Research Foundation, Singapore, Singapore
e-mail: low_teck_seng@nrf.gov.sg

are offering unprecedented opportunities for increased energy efficiency savings while empowering users to manage and optimise their assets and processes.

3. **Do you think digitalisation has significant potential for reducing CO_2 emissions in Singapore? If yes, what are the areas of interest and by how much?**

Digitalisation is a cross-cutting tool that will help us manage and optimise our energy supply and demand to reduce energy consumption. It also serves as an enabler to facilitate the large-scale deployment of new energy technologies into our existing systems, such as electric vehicles (EVs) and solar photovoltaics (PV). With digitalisation, we are able to shift towards a smarter, decarbonised and decentralised energy system, in tandem with the rise of renewables and distributed generation, market deregulation and technological developments. This would involve integrating renewable sources through smart inverters, deploying smart meters and sensors, and introducing energy management systems—all of which would help to decarbonise our electricity grid and reduce CO_2 emissions in Singapore.

4. **What R&D and education investments are necessary for Singapore in order to contribute to reducing CO_2 emissions and cope with the impact of future digitalisation?**

NRF has embarked on the Energy Grid 2.0 R&D programme, which aims to develop the next-generation grid system that will transform how energy is managed. By consolidating gas, solar and thermal energy into a single intelligent network that is more efficient, sustainable and resilient, we can address Singapore's unique energy challenges. This programme builds on our past investments in smart meters, grid storage and solar PV, as well as various energy efficiency and demand management solutions.

5. **If there is one problem that you could solve with the use of AI, what would it be?**

How to better manage Singapore's consumption and production of energy in order to tackle climate change concerns. A research study[1] published by PwC UK last year estimated that AI could help to reduce greenhouse gas emissions by 4%, boost global GDP by up to US$5 trillion and create 38 million jobs by 2030. There is enormous economic and environmental potential for AI adoption in the agriculture, water, energy and transport sectors, which we can and must pursue in Singapore.

---

[1] How AI can enable a sustainable future.

Teck Seng Low is the CEO of the National Research Foundation, Singapore. He was previously the Managing Director of A*STAR, the founding principal of Republic Polytechnic, as well as the Dean of the Faculty of Engineering in the National University of Singapore. Professor Low was awarded the National Science and Technology Medal in 2004 and the Public Administration Medal (Gold) by the President of Singapore for his outstanding contributions to the development of technical education and the management of science and technology for the nation. In 2016, he was conferred the Order of the Legion of Honour with the grade of Knight by the French Government in Paris.

Chapter 30
Insights: Pollutant to Feedstock

Volker Sick

1. **What are the global CO$_2$ initiative's approaches to reducing CO$_2$ emissions?**

Beyond the urgent need to massively reduce CO$_2$ emissions, we must also address legacy emissions in the atmosphere and continuing emissions from so-called hard-to-abate sectors. Our focus is on developing CO$_2$ into a valuable feedstock to make products that cannot be made without carbon and to commercialise products that can permanently remove carbon from the atmosphere. These efforts immediately contribute to decarbonisation, which does not mean that we have to stop using carbon; rather, we must stop adding new carbon into the atmosphere. Thus, creating a circular economy with carbon-based products made from captured CO$_2$ will be an essential element in climate action.

2. **What role does digitalisation play in CO$_2$ utilisation today and in the future?**

Implementing a carbon economy built around a CO$_2$ feedstock requires building up entirely new supply and value chains. This new carbon economy can be structured around current technological infrastructure to improve efficiency and real-time communication. In designing and operating new industrial processes, digital tools can optimise additional factors, including globally minimised CO$_2$ emissions, social equity and global resources management. None of that will work without the fullest deployment of digitalisation tools across the CO$_2$ utilisation field.

3. **Do you think digitalisation has significant potential for reducing CO$_2$ emissions in the US in the future?**

There is no doubt in my mind that continuing enhancements and implementation of digitalisation will be a technological enabler to optimise systems for minimal CO$_2$

V. Sick (✉)
University of Michigan, Ann Arbor, USA
e-mail: vsick@umich.edu

© Springer Nature Switzerland AG 2022   235
O. Inderwildi and M. Kraft (eds.), *Intelligent Decarbonisation*, Lecture Notes
in Energy 86, https://doi.org/10.1007/978-3-030-86215-2_30

emissions. The key is that only with full consideration of any action's impact on the system response, we will ultimately achieve net-zero emissions. Technological tools are needed for this rapid systems evaluation. Beyond optimised technology operation, fully integrated digital systems will enable us to transparently track emissions, to record certification of carbon credits, carbon dividends, tax obligations and other financial instruments essential to manage ambitious environmental goals.

4.   Why are digitalisation and artificial intelligence critical in addressing climate change?

Digital tools are already being used to react to the worst impacts of climate change. Big data and AI are being used to respond to fires, floods and other climate disasters. They have shown their potential to save lives and minimise economic losses. However, these tools must be used for more than adaptation; climate regeneration industries must embrace digital tools in the toolkit needed to fight climate change. Among other means, artificial intelligence and digitalisation tools must be brought to a level of performance and reliability to accurately predict the system-level impact of proposed climate action in all aspects: climate-related, financial or societal.

Volker Sick leads the Global CO_2 Initiative at the University of Michigan that seeks to get CO_2 capture and use recognised and implemented as a mainstream climate solution. He received awards for teaching, research and service, including the President's Award for Distinguished Service in International Education, the Combustion Institute Silver Medal and the SAE International Leadership Citation. He is a Fellow of SAE International and the Combustion Institute. Professor Sick holds three degrees in Chemistry and Physical Chemistry from the University of Heidelberg, Germany.

Chapter 31
Insights: Digitalisation and China

Donghan Jin

1. What are China's approaches to sustainable development and what role does higher education play in this context?

In China's 14th Five-Year Plan (2021–2025) for National Economic and Social Development and the Long-Range Objectives Through the Year 2035, detailed approaches are proposed for sustainable development: (1) to promote green and low-carbon development and reduce carbon emissions; (2) to improve environmental quality and promote environmental protection; (3) to improve ecosystem quality and stability; and (4) to improve resources and energy utilisation efficiency.

To support the national strategy, higher education must take proactive action to change our attitude from responding to change to leading change, as well as educating engineers and leaders to build our future world. To this end the Emerging Engineering Education (3E) Plan was initiated by the Chinese Ministry of Education in 2016. The 3E plan aims at educating future engineering innovators and calls for a new round of engineering education reform to transform the traditional engineering education mode.

Tianjin University (TJU) took a leadership position in this process, proposing the 3E Roadmap in 2018 and subsequently launching the International Alliance of Emerging Engineering Education (IAEEE), consisting of 56 universities from China, France, Ireland, Singapore, Thailand and the US, to address the challenges faced by global engineering education communities. TJU has become the centre of the 3E movement in China and the Ministry of Education has formally established a National Innovation Centre for 3E at TJU.

D. Jin (✉)
Tianjin University, Tianjin, China
e-mail: president@tju.edu.cn

© Springer Nature Switzerland AG 2022   237
O. Inderwildi and M. Kraft (eds.), *Intelligent Decarbonisation*, Lecture Notes in Energy 86, https://doi.org/10.1007/978-3-030-86215-2_31

2. **Do you think digitalisation has significant potential for reducing carbon emissions in China? If yes, what are the areas of interest and by how much?**

Manufacturing is a pillar industry of China and contributes over 30% of its total carbon emissions, with major contributors including coal, chemical and transportation. Digitalisation can improve production efficiency and quality, thus reducing the cost and energy consumption. In China, many manufacturing enterprises have built digital and intelligent factories. The China Shipbuilding Industry Corporation's digital shipbuilding workshop is equipped with a variety of welding robots to replace the manual work for sequence welding, which increases manufacturing efficiency by 30% and reduces energy consumption by 10.8%. With the help of artificial intelligence (AI) technologies, the China Aerospace Science and Industry Corporation has achieved a 70% comprehensive utilisation rate of equipment in the whole workshop, increased the production efficiency by 30% and reduced the manpower demand by more than 50%. The reduced manpower demand workshop can better guarantee safety in the post-pandemic era. Furthermore, AI technologies can make household appliances, vehicles, internal combustion engines and other products safer, cheaper, more reliable and more environmentally friendly. In addition, the traditional maintenance processes of many industrial products waste much energy due to the lack of quantitative estimation of the degradation level. Future big data and digital twinning technologies are expected to realise real-time detection of products and reduce energy consumption and carbon emissions.

3. **If there was one problem that you could solve with the use of AI, what would it be?**

One of the main targets of "Made in China 2025" is to achieve the intelligent transformation of the manufacturing industry. In my view, AI technologies will help such a transformation in three ways. The first is the intelligentisation of the product itself, such as an intelligent engine that is more efficient, more environmentally friendly, more reliable and safer. The second is the intelligentisation of the production process, which leads to high-quality, high-efficiency and low-cost manufacturing. The last is the intelligentisation of after-sales service, for example, an intelligent car that can alert the driver that maintenance is required instead of relying on a fixed period of time or driving distance. In China, accelerating the development of intelligent manufacturing is an important measure to promote supply-side structural reform, and to upgrade China from a manufacturer of quantity to one of quality.

Professor Donghan Jin is the President of Tianjin University and an Academician of the Chinese Academy of Engineering. He is also the President of the International Council on Combustion Engines (CIMAC) and the Chinese Society for Internal Combustion Engines (CSICE). Professor Jin obtained his Ph.D. degree from the China Ship Research and Development Academy in 1989. In July 2015, he was appointed as the President of Shanghai University. Previously, he served as the President and Chief Engineer of the Shanghai Marine Diesel Engine Research Institute for more than 11 years. Professor Jin has been engaged in the research and development of the Stirling engine and its power system. He has published a monograph on the Stirling engine and a book on artificial intelligence. He has won one Outstanding Prize and one First Prize of the National Science and Technology Progress Awards, and three first prizes at the provincial and ministerial level. He has also received more than 20 national and provincial honours, including the Science and Technology Progress Award by Ho Leung Ho Lee Foundation and the Hero Award of Shanghai Science and Technology, the top science prize of Shanghai.

Part VI
Conclusions

Chapter 32
Synthesis

Oliver Inderwildi and Markus Kraft

Abstract Mitigating climate change will require significant efforts in all domains ranging from technological progress to societal transformation and from economic restructuring to political guidance. This compendium clearly elaborates that digital progress will be a key tool that facilitates the transition to net-zero emissions. Within this chapter, the key trains of thought and key technologies described by contributors are synthesized to create a coherent picture of the key technologies, vital trends and approaches necessary to support the transition of the global economic system. The synthesis clearly demonstrates that digital technologies not only improve and augment the operation of current systems responsible for greenhouse-gas emissions, but also are a vital planning and decision support tool that aids the evolution of systems to be inherently more sustainable. Moreover, key technologies that positively affect all domains are determined, which implies distinct technological areas that policy makers should support in order to achieve broad emission reduction and economic benefits.

O. Inderwildi (✉) · M. Kraft
CARES, Cambridge Centre for Advanced Research and Education in Singapore, 1 Create Way, CREATE Tower, #05–05, Singapore 138602, Singapore
e-mail: oliver.inderwildi@essec.edu; oliver.inderwildi@scnat.ch

Swiss Academy of Sciences, Forum for Global & Climate Change, Maison des Académies, Laupenstrasse 7, 3008 Berne, Switzerland
e-mail: mk306@cam.ac.uk

M. Kraft
Department of Chemical Engineering and Biotechnology, University of Cambridge, Philippa Fawcett Drive, Cambridge CB3 0AS, United Kingdom

School of Chemical and Biomedical Engineering, Nanyang Technological University, 62 Nanyang Drive, Singapore 637459, Singapore

© Springer Nature Switzerland AG 2022
O. Inderwildi and M. Kraft (eds.), *Intelligent Decarbonisation*, Lecture Notes in Energy 86, https://doi.org/10.1007/978-3-030-86215-2_32

32.1 Introduction

Humanity is at a crossroads as it faces two existential risks: firstly, climate change caused by anthropogenic greenhouse gas (GHG) emissions could lead to an uninhabitable atmosphere. Secondly, the advent of human-like or artificial general intelligence (AGI) could—analogue to uncontrolled climatic change—lead to the extinction of humankind as it would be subject to the goodwill of a higher intelligence. Overcoming these risks is not optional if humanity is to survive. Fortunately, both risks also give rise to unprecedented opportunities: restructuring the global economy to net zero emissions would mitigate and reverse climate change and inevitably lead to a healthier, more liveable world, while AI could be utilised to address global problems that the human mind is not capable of solving. This book brings together leading thinkers from all walks of life to discuss how AI and digital technologies could support the UN campaign Race to Zero and mitigate or even reverse climate change. In this chapter, the findings are synthesised and common threads are established.

32.2 Methods

Cyber-Physical Systems: The impact that can already be seen from digital technologies stems not only from digitalisation of specific processes, but also from so-called cyber-physical systems (CPS). These are defined as high-level orchestrations of various cutting-edge digital technologies that create a digital representation of the physical world. This digital representation, often also referred to as a Digital Twin, enables the enhanced operation and control of the physical system using advanced optimisation tools. Chapter 2 outlines the economic and environmental benefits delivered by CPS and illustrates that development is only in its advent. In Part III of this book, many examples of CPS and their impact are described and give an outlook on what is to come. A holistic framework is provided in Chap. 15, where a four-pillar framework for the successful deployment of digital technologies for energy provision is outlined. Firstly, sensing and data fusion will provide valuable information on energy systems; the capability was facilitated by recent advances in Internet of Things, low-cost cloud storage and internet bandwidth. Secondly, prediction of supply and demand by algorithms based on deep learning will play a critically important role. Thirdly, as set out in a number of chapters, AI-supported probabilistic models can create differing scenarios which act as excellent decision support tools. Last but not least, the fourth pillar is the actuation of improvements established in the Digital Twin via CPS. It can be concluded that seminal developments in specific digital competencies led to ground-breaking real-world benefits such as cost and emissions savings via technology orchestration. Key to the success of digital competencies is the efficient exchange of data, with the physical world via CPS and within the virtual representation, *i.e.,* the Digital Twin.

Knowledge Graphs and Avatar: In order to facilitate the exchange of data, *i.e.* within the Digital Twin and with the physical system via CPS, novel concepts such as dynamic Knowledge Graphs (dKG) are introduced. Chapter 4 sets out how operational expenditure (OpEx) as well as emissions can be reduced using Intelligent Control Strategies and underpins their claim with examples from the chemicals industry. The dKG "The World Avatar" facilitates the sharing of data between processes, which allows alignment of such processes and minimises overall energy use, hence reducing emissions and OpEx; facilitated co-generation of heat and electricity, for instance, reduced energy use and therefore emissions by 16% via an efficient power/heat dispatch strategy. Moreover, peak-demand scenarios could be avoided by an energy demand side management framework based on a dKG approach that utilises combined forecasts for renewable energy production, energy demand for three sectors and facilitated energy trading via Blockchain technology. The efficient exchange of data via the dKG encouraged uptake of renewable energy sources (RES) and enhanced the efficiency of the overall system leading to emission reduction up to 46% compared to no-RES scenario. While digital technologies and their orchestration already provide significant benefits and will play a key role in the decarbonisation of the globalised economy, artificial intelligence is likely to hypercharge these endeavours.

Artificial Intelligence: At specific tasks, AI has beaten all expectations by a significant margin; the dethroning of chess and Go grandmasters has clearly shown us the power of machine intelligence. However, the human mind is still more versatile as chess computers cannot solve everyday tasks—at the moment AI is still specific or narrow. Nevertheless, the skillset of AI will broaden over the coming years and its success in one area will lead to success in another, for instance optimisations from server cooling might be applied to the energy management of office buildings. AI can handle data sets that are beyond the abilities of humans, even using the most advanced current techniques; the availability of large data sets plus the specific insights provided by current AI delivers new insights, which in turn improve the strategies and decisions made by humans. According to David Rolnick of McGill University (Chap. 5), AI is the ultimate decision support tool for human decision makers. This claim is supported in every chapter and interview of Part III, all of which highlight the improved planning and decision support provided by AI-enhanced CPS.

The development does not stop here: quantum computing will supercharge the transition from narrow to artificial general intelligence (AGI), the creation of which will inevitably lead to singularity (see Chap. 3). Technological singularity will leave us with the so-called AI control problem: how can humans control an intelligence far greater than their own? In addition, such an intelligence is likely to be conscious; such a consciousness is unlikely to resemble that of humans, but machines will be aware of themselves in some respect. This clearly philosophical question is raised in Chap. 22, which scrutinises whether machine consciousness should be prevented as from this inflection point, machines are likely to interfere with humans' affairs. This compendium therefore concludes that significant investments in AI safety research are required and these investments should be made now, because once a conscious AGI is created, it would be too late. Owing to the potential benefits and risks of

AI, it is critical to channel knowledge, create safeguards and inform the public. Peter HO Teck-Hua, Provost of the National University of Singapore and Executive Chairman of AI Singapore, lays out Singapore's highly successful AI Singapore initiative that channels the AI knowledge of research institutions, the start-up scene and corporations to deliver benefits through AI. The initiative also has the key role of informing the public and raising awareness of the ethical and governance issues of AI. Addressing these issues is critical for the safe implementation of AI and this is discussed in Part IV.

Blockchain Technology: Distributed ledger technology is yet another means with the potential to optimise many sectors while delivering emissions savings. Chapter 7 explains how blockchain technology—the most applied distributed ledger—can simplify complex transaction processes by eliminating the need for intermediaries; this simplification reduces energy use and this simultaneously reduces cost and carbon footprint. In the case of supply chain optimisation this could lead to emissions savings of up to 50%, while in the energy sector savings well beyond 30% are on the cards, according to Chap. 7. In the energy sector, the most popular usage of blockchain is decentralised energy trading, which allows the energy system to be more responsive. It is noted, however, that the enhanced trading must be combined with augmented demand-side predictions—most likely stemming from AI—which is also discussed in Chaps. 4 and 12. Analogue to the increased transparency provided by smart grid technology, Chap. 7 subsumes that in both energy markets and supply chains, blockchain technology will provide transparency and hence accountability. This will be critical for emissions markets as it reduces the risk of fraud and double (ac)counting leading to a higher acceptance of this useful tool for holistic emission reduction. The question that remains is with whom the responsibility lies when an algorithm makes decisions. From a legal perspective, Chap. 22 argues that smart contracts are merely automated contracts and hence neither take decisions nor assume responsibility. In this case, the legal entities behind the smart contract assume responsibility, because they agreed on the goals—even though on a meta-level—while the smart contract merely fulfils these goals in an automated manner. With regard to the cost-benefit it is highlighted that blockchains are computationally expensive, an issue that is discussed later in the book.

32.3 Sectors

Processes and Chemicals: These technologies are clearly interesting and have great potential, but do they already have a significant impact and what might their impact be in the future? This is the critical question that experts from academia and industry elaborate on in Part III with distinct examples and professional estimates for future gains.

Under the concept of Industry 4.0 the producing industry is already utilising cyber-physical production systems (CPPS) with serious implications for production efficiency, costs and emissions. According to Chap. 8, CPPS unfold their full potential in complex production situations where there is sufficient potential to harness co-benefits between different processes. However, the most promising CPPS applications are on the systems level where integration of larger industrial entities can deliver significant savings by synchronising their energetic needs. This is a key criterion of Industry 4.0 and already implemented by e.g. the Verbund concept by BASF (see Chap. 9). The Verbund has already delivered the company significant economic benefits and simultaneously reduced the environmental footprint of their products, vide infra. Thiede also anticipates further interesting levers such as, for instance, the synchronisation of production processes with external conditions such as the weather to minimise the energetic needs of such processes. DeepMind's optimisation of Google's servers is a first example of such a synchronisation (see Chap. 3). Based on an assessment of the literature as well as the authors' own research, Chap. 8 estimates the energy improvement potentials in plant operation to be over 15%.

This view is supported by the experiences from the above-mentioned Verbund concept: using this approach, BASF was able to reduce costs by over €1bn while abating more than 6Mt of $CO_{2(eq.)}$ facilitated by process integration. According to BASF's experience (Chap. 9) the concept of Industry 4.0 will significantly enhance the benefits of process integration. Moreover, the chemicals industry is asset heavy with long lifetimes; the former requires large investments while the latter creates inertia. The Industry 4.0 approach enables companies to optimise existing assets and thereby minimises capital expenditure. Chapter 10 supports this view and recommends fast action due to the above-mentioned asset inertia in this industry; poor investments will carry into the future and lock the industry into suboptimal processes. Quick wins through predictive analytics, resilient network structures and self-organisation for this highly complex industry are nonetheless possible under Industry 4.0 but also facilitate the development of green business cases. Deductions from industrial experience at Siemens give similar estimates for the reduction potential in chemical industry, namely 17% in energy consumption and hence emissions. Knowledge transfer between sectors will be facilitated by digital technologies, *i.e.*, learnings from one sector will be applied to another to deliver emissions savings (Chap. 11).

Electricity: The sector that is key to all decarbonisation endeavours and which will provide spillover effects is electricity provision and here the smart grid approach already provides many benefits. This approach increases the infrastructure's sustainability and utility through digital optimisation of the system's operation and intelligent planning of its evolution. Smart grids not only include self-regulating and self-healing attributes allowing real-time response but also avoid supply and demand peaks, which can lead to electricity waste or high-carbon supply. Adaptricity's experience with an electricity provider from Liechtenstein has shown that the real-time insights provided by continuous monitoring can reduce line losses up to 10%, reducing emissions and operating costs (see Chap. 12). This chapter anticipates that the hosting capacity for renewables can be increased by 20% due to intelligent distribution of intermittent over-capacity to *e.g.*, electric vehicles or heat pumps. Moreover, low-carbon electricity provision will be critical to decarbonise the most important resource for humanity, water.

Water: Climate change will affect the availability of fresh water and as a consequence, energy-intense processes such as desalination will be critical to ensure water security. Chapter 13 discretises critical factors that define the energy and emission footprint of water and discusses how AI and digital technologies can reduce both. It is concluded that the energy savings of water provision can reach 16% using digital technologies. Moreover, the smart utilisation of wastewater treatment by-products can provide production plants with carbon neutrality, while the emissions from desalination plants can be drastically reduced by utilising *e.g.,* solar power.

Urban Systems: Digital technologies can also provide benefits on a smaller scale, according to the experience of a local German utilities provider who integrated their waste utilisation with energy provision (Chap. 14). Energy management systems facilitate the reduction of energy needs of the utilities themselves, which further aids the decarbonisation of their provision. Local utilities and grid operators will face yet another challenge—the uptake of electric mobility on a large scale. Electric vehicles (EV) will increase the demand for electricity and will simultaneously provide a distributed storage network through their batteries. These two points will make smart grids and intelligent utility provision even more important over the coming years (see Chaps. 12 and 18 on smart grids and transport respectively). The integration of urban energy networks is also discussed in Chap. 15, where it is estimated that smart operation of urban energy networks could reduce costs and emissions by 10 to 30% with very short payback periods for investments in digital support technologies. The chapter assesses a pilot project of a District Energy Scheme in the UK in which heating bills could be lowered by 10%, while GHG emission could be reduced by a staggering 60%. When the system of reference is enlarged, for instance by coordination of different districts, emissions can be reduced by up to 77% and operational expenditure by 30%. For the United Kingdom, synchronising heat, electricity and transport to save costs and emissions is key, however, sub-tropical and tropical cities face yet another problem. Chapter 17 sets out the issue of urban heat islands (UHI) caused by local CO_2 emissions in agglomerations in sub-tropical climates. The comprehensive analyses of UHI effects require not a single digital approach but an orchestration of advanced, coupled models as in the Digital Urban Climate Twin (DUCT): firstly, a realistic digital representation of all relevant features of a certain settlement and secondly, an expandable network of coupled models in order to create simulations of desirable scenarios for the future city. For Singapore, which is used as a representative example, this will entail the assessment of potential pathways for decarbonisation including a cross-sectoral view of the benefits, trade-offs and synergies of different decarbonisation strategies. Again, this is a clear example that the first applications of digital orchestrations combined with AI methods will be advanced planning and decision support. For sub-tropical and tropical cities dependent on energy provision by fossil fuel, there will be a virtuous circle as the reduction in emissions will help to reduce the urban heat, which in turn reduces the need for cooling and air conditioning and reduces emissions. This virtuous circle will make these cities more liveable and resilient due to, for instance, the improving air quality and noise reduction. Schmitt and co-workers estimate that if these technologies were implemented, Singapore could reduce its emissions by more than 50% by 2040 and simultaneously reduce its utility budget by several billion SG$ per annum. In addition, managing and planning buildings will be key to a sustainable future. Singapore's JTC Corporation reports that infrastructure deployment can be optimised throughout the lifecycle using digital technologies, starting with the design and planning of efficient estates to increasing construction productivity and reducing production waste, and finally monitoring and optimising energy needs of operating buildings. AI will hence play a critical role from planning to operation and upgrading delivering key savings in cost and emissions.

Transport: The next important sector to decarbonise is transport as it is responsible for roughly a quarter of global emissions. Recent developments in electric vehicle (EV) uptake clearly indicate that individual mobility will turn mainly electric within this decade. This will increase the demand for electricity, but also will make the grid more resilient due to the distributed storage it introduces; the latter is a key prerequisite for increasing the share of intermittent renewables feeding the grid. Chapter 15 presents a case study in which the shift of excess PV production to EV charging led to a 10% reduction of EV related emissions, while in another a 50% reduction of peak electricity demand using a smart building approach allows for additional uptake of intermittent renewable generation. Such an integration will successively make electricity the new primary energy. Not only individual transportation has to be optimised, however—transport of humans and goods as a system must be integrated, analogue to the integration of chemical processes, vide supra. In transport systems, start and end point, purpose, mode choice and the motivation for the choice of mode are critically important. Here, AI has a key role to play in supporting human decision makers: first, by validating digital twin models of the transport system; and second, by estimating the response of consumers to decarbonisation policy interventions for sustainable mobility (Chap. 18). This substantiates this theory with case studies for successful pilot programmes in Herrenberg, Zurich and Cambridge (UK). AI-supported automation does not guarantee a reduction in emissions from transport, but Chap. 18 accounts that transport emissions could be 45% lower through AI-supported automation in a best-case scenario. Realising this impact requires strong policy to steer the deployment of AI to serve both society and the environment, avoiding any unintended consequences.

Markets: Hamacher of TUM also argues that the coupling of heat, power and transport could reign in a new era in which electricity is the new primary energy (Chap. 19). However, this transformation should be market-driven in order to find cost-optimal solutions. To achieve this, information asymmetry has to be reduced, which is best realised via an Open Information Platform that informs market participants and overrides information asymmetry. Open Data Platforms, as advocated in Chap. 19, can measure the performance of a certain policy on a daily basis and thereby reassesses and improves policy decisions. This enables a market-driven energy transformation leapfrogging a tedious trial-and-error process.

Cost-Benefit of Digital Solutions: Another critical conclusion from this compendium is the consideration of the economic and environmental cost-benefit of CPS and AI deployment, i.e., do the benefits outweigh the costs for our society and our environment? There is a clear consensus throughout this book that not only the environmental but also economic cost-benefits ratio should be assessed, while all authors agree that in most cases the benefits clearly outweigh the costs. The return on investment for digital solutions is usually relatively short (Chaps. 3, 8, 9 and 15), while they allow the optimisation of otherwise inert systems—i.e. assets with long lifetimes (Chaps. 9, 10, 12, 16, 18 and 28). It can therefore be concluded that digital solutions are favourable in monetary and environmental terms. Moreover, the ability of CPS to reduce the emissions from assets with long lifetimes reduces the cradle-to-grave emissions from these industries; without CPS, assets would need to be replaced prematurely to save emissions while the emissions embedded in the asset (created by manufacturing them) would essentially not be fully utilised; CPS and AI enhance the utility of current infrastructure (Chap. 9), another key factor to be considered for the costly transition to a net-zero economy—current infrastructure has to be used optimally over the entire lifecycle. The only area in which this is not true at the moment is blockchain technology (see Chap. 7), but many innovative approaches are under consideration that will reduce the energetic need for blockchain mining such as the proof-of-stake concept. In addition, blockchain technology can help electricity markets to reduce emissions; since electricity is the main input for mining, yet a *positive feedback loop* is created.

Capital vs Operational Expenditure: From the preceding paragraphs it becomes apparent that CPS and AI can and will optimise processes, which lowers operational expenditure by reducing energy needs and consequently emissions. Yet another clear deduction from this compendium is that digital systems will reduce capital expenditure as well. Within Digital Twins, possible scenarios can be planned and tested in order to select the variant that provides the highest emissions savings at the lowest cost (Chaps. 4, 10, 12, 15, 17, 18). Capital expenditure for the producing industry will be reduced by minimising unnecessary or superfluous infrastructure and machinery; the reduced expenditure clearly goes hand-in-hand with a reduced environmental burden and consequently the emission reduction can reach 25% (Chap. 8). In the case of electricity grids, intelligent grid management will support sophisticated decision making and lead to advanced asset management by estimating, for instance, infrastructure lifetimes based on usage or lifetime costs. Managing these infrastructure assets using optimised grid investment strategies hence optimises capital expenditure in the long-run (Chap. 12). In transport, the mode choice of consumers is influenced by enabling infrastructure. Also in this case, AI can support the planning of this via scenarios within the Digital Twin and the scenario that provides the most efficient and cost-effective transport system with the least environmental impact is implemented. The same is true for the dKG "The World Avatar", which also provides capability scenarios planning using the parallel world concept, *e.g.*, an energy storage system can be planned so that it delivers the optimal cost-emissions savings by analysing scenarios. It is very likely that AI will also facilitate the scenario generation and provide human operators with unconventional scenarios that deliver a superior cost-benefit ratio. In the case of chemical process, for instance, the dKG determined that with a certain control strategy, a lower-rated transformer would be sufficient resulting in €40,000 Capital expenditure savings. In the case of policy interventions such as a progressive carbon tax, the AI-supported scenario planning can moreover create roadmaps on the transformation of the energy supply system (Chap. 4). This book therefore clearly illustrates that AI and CPS will be critical decision support tools for the decarbonisation of the economic system.

32.4 Legal and Governance

All the benefits of digital technologies and AI described in Part I, II and III come with distinct risks and therefore Part IV addresses the legal, ethical and governance implications for these technologies and suggests distinct solutions. What are the potential downsides of AI? Firstly, AI enhances the possibility and severity of cyber-attacks, a phenomenon that can already be observed in global attacks by hackers (Chap. 3). Secondly, there will be structural effects imposed by AI due to its consequences on human affairs ranging from commercial dealings to societal interactions. These effects may include serious displacements of labour, privacy concerns and ultimately the meaning of our lives, as AI has the potential to perform many tasks for us with minimal action required on the human side (Chap. 22). Thirdly, AGI poses an existential threat to humanity which is explained in Chap. 3. Last but not least, the rearrangement of the global power balance due to the race for AI leadership; countries with the most advanced AI capabilities are likely to be the dominating forces of the future (Chaps. 3 and 22). AI, however, should first and foremost serve humanity and the planet and in order to ensure this, novel legal and governance approaches are required that will lead to goal alignment between AGI and human goals. Due to AI's potential for efficiency increases in almost all affairs of life, its implementation is likely to be driven by the forces of economics and a deep-rooted belief in technological progress (Chaps. 1 and 3).

Two distinct approaches for the governance of AI are proposed in this book: a bottom-up network governance as well as a more centralised meta-governance that urge environment- and human-centric AI principles (Chap. 20). What are some specific security issues? It is argued that a smart city approach is at the same time a rather perfect surveillance method (Chap. 21 and 23). This implies that, due to privacy laws and regulations, the approach will not meet its full potential unless solutions are found that provide benefits from the smart city approach without impinging on data security; hence, the legal and governance system has enforced a healthy balance between benefits and risks. Chapter 22 substantiates this: The advances in AI will inevitably demand new laws and regulations, because these novel technical systems cannot be fully influenced or understood by their users. The legal concept of responsibility, however, is so intrinsically linked to human beings that it seems illusory to hold a machine liable for breaching law. Chapter 22 proposes to shift from contemplating causality to considering final goal-setting in order to find categories for assigning responsibility; goal- and the rule-setting that limit actions of an intelligent machine have to be assessed. Chapter 22 predicts that legal obligations will shift—at least partially—from operators to manufacturers and the latter will emphasise preventative approaches in order to minimize their risk. The legal liability for machines will therefore still lie with humans for the foreseeable future, while a transferral of liability from the operator to the manufacturer will take place with increasing sophistication of the machines' AI ability. Existing regulations will not prove to be ideal in every respect and hence the need for regulatory and legal adaptation will certainly arise. Chapter 23 also observes that technology and the law regulating it widens as the progress in—especially digital—technology outpaces the progress in law-making. The contribution proposes principle-based and technology-agnostic laws and regulations that are purposive, which increases their adaptiveness and reduces their redundancy. Only this way, the legislative process can keep up with technology developments. In general, ethical governance frameworks must be established and these have to be met irrespective of the technology. Advanced technologies require built-in ethics, especially in the areas of privacy and security where there is an imperative for identifying and mitigating bias introduced by AI (Chap. 23). From Part IV it becomes apparent that both legal systems and global governance are ill-equipped to deal with the emergence of AI and moreover, the speed of adaptation will likely outpace the speed of legislative processes; innovative ideas for adaptive, ethical legislation are therefore key for ensuring that AI and digital technologies will deliver benefits while minimizing the risk for humans.

32.5   Big Picture

In addition to these insights from academic and industrial researchers as well as the ethical considerations from scholars, it is important to consider the bigger picture. What are institutions doing to reduce carbon emissions and are digital technologies as well as AI part of their strategy? The Vice-Chancellor of the University of Cambridge, Stephen Toope, lays out how the university coordinates its research and policy endeavours via the Cambridge Zero initiative. According to the science manager, digital technologies will be indispensable for the transition to net zero emissions. However, he sees risks—analogue to those observed in Part III—with regards to bias introduction and intrusive surveillance. This is the reason the university introduced a Masters programme for responsible AI (Chap. 24). Christian Thomsen, President of the Technical University of Berlin, explains how Germany sees the same opportunity and channels financial resources into National AI centres in order to foster and incubate AI technology and realise its potential (Chap. 25). Professor Donghan Jin, President of Tianjin University and Editor-in-Chief of the journal "Energy and AI" describes how China plans to benefit from cyber-physical systems and AI. He highlights how the Emerging Engineering Education (3E) plan, issued by the Chinese Ministry of Education, will produce the expertise necessary for the implementation of new AI technologies. (Chap. 31). Paul Monks, the Chief Scientific Advisor at the UK's Department for Business, Energy and Industrial Strategy dives into recent policy developments in the UK. The UK has set up £3 billion National Skills Fund that includes investments for upskilling and retraining so that the human resources for the transition to Net Zero are available. The Science Advisor sees a flexible power system as a key prerequisite for its decarbonization and digital technologies as essential to provide such a flexible system (Chap. 27). Lord Julian Hunt, member of the British House of Lords, reminds us that AI has to be explainable in order to be adapted on a large scale, that is humans have to be able to reconstruct how an AI came to its decision. A critical factor mentioned throughout this book. Moreover, his Lordship sees for instance an issue in the perception of nuclear power as it is meanwhile a very safe technology that could provide true low-carbon hydrogen (Chap. 28). Low Teck Seng of Singapore's National Research Foundation is also optimistic about the impact of digital technologies and AI. In line with Baumgartner & Ulbig (Chap. 12), he sees intelligent grid management as key to the large-scale deployment of photovoltaics and electric vehicle, only with intelligent management of the grid can this be achieved. Moreover, the research administrator sees opportunities in agriculture, water provision and transport—all of these are critical sectors for the city-state (Chap. 29). Volker Sick of the University of Michigan's Global CO_2 Initiative is convinced that artificial intelligence can predict the systems level impact of climate policy measures and moreover, aid the local abatement and utilization of CO_2 (Chap. 30).

From Part V it can clearly be seen that leading academic managers, government advisors and research administrators see enormous potential in the technologies discussed herein and their potential to mitigate climate change and fulfil net-zero obligation. This claim is clearly supported by the fact that very significant financial resources are channelled towards research and development, upskilling and consolidation of knowledge in this area. Moreover, the establishment of centres of excellence, dedicated degrees and innovation hubs that aim at tackling climate change and decarbonization with digital technology underlines the issue and the timeliness of this book. Our overall conclusions for this input will be summarized in the subsequent chapter.

Chapter 33
Conclusions

Oliver Inderwildi and Markus Kraft

Abstract In this concluding chapter of this compendium, the essence of all contributions is presented. It will be shown that digital technologies, especially when combined with artificial-intelligence capabilities, have the potential to fully integrate economic sectors and, in combination with electrification, steer the globalised economy towards net-zero emissions. The transition is hence key to mitigating climate change and the existential risk it poses; nevertheless, the digital transformation itself bears existential risks. Owing to this, this compendium is concluded with key recommendations to maximise the economic and environmental utility of digital technologies and artificial intelligence while minimizing the risks these technologies inherently introduce.

This compendium illustrates that digital technologies and AI are critical innovation providing excellent new tools that will aid the achievement of net-zero emission targets and consequently combat climatic change.

At the core of net-zero strategies is the electrification of economic sectors ranging from transport to industry and hence electricity is designated to be the dominant form of energy provision. Industrial and academic researchers as well as managers from the

O. Inderwildi · M. Kraft (✉)
CARES, Cambridge Centre for Advanced Research and Education in Singapore, 1 Create Way, CREATE Tower, #05-05, Singapore 138602, Singapore
e-mail: mk306@cam.ac.uk

O. Inderwildi
e-mail: oliver.inderwildi@scnat.ch

O. Inderwildi
Swiss Academy of Sciences, Forum for Global & Climate Change, Maison des Académies, Laupenstrasse 7, 3008 Berne, Switzerland

M. Kraft
Department of Chemical Engineering and Biotechnology, University of Cambridge, Philippa Fawcett Drive, Cambridge CB30AS, UK

School of Chemical and Biomedical Engineering, Nanyang Technological University, 62 Nanyang Drive, Singapore 637459, Singapore

© Springer Nature Switzerland AG 2022   255
O. Inderwildi and M. Kraft (eds.), *Intelligent Decarbonisation*, Lecture Notes in Energy 86, https://doi.org/10.1007/978-3-030-86215-2_33

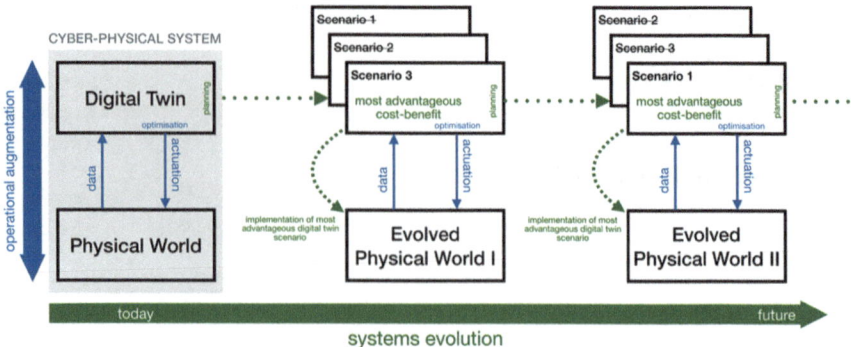

Fig. 33.1 Cyber-physical systems (CPS) not only improve existing systems, but are critical to planning a systems evolution which provides economic and environmental benefits

private sector provide case studies to illustrate that all sectors rely not only on internal efficiency improvements, but also on the provision of low-carbon electricity. This drive towards electrification will couple many sectors that either supply or demand energy. This coupling provides many benefits in terms of efficiency improvements and the harnessing of co-benefits, but also poses issues with regard to the management of such a complex and integrated system. Digital technologies—especially when connected to AI capabilities—are predestined and key to the enhanced management and steering of a fully integrated, electrified economic system.

Operational augmentation: Cyber-physical systems and the orchestration of digital technology are already very successfully applied in many sectors to improve internal efficiency. Key to these improvements is the so-called digital twin, a digital representation of the physical world in which intelligent computer algorithms optimise the physical system based on data collected from its real-world equivalent. The actual benefits are brought by the CPS as it provides the digital twin with the data required for process augmentation and is paramount for the actuation of certain improvements in the physical world (left-hand side, Fig. 33.1). Throughout this book, illustrative examples are provided of the potential reductions in emissions and operational expenditure of CPS. The contributions all agree that digital technologies have short payback periods with high returns on investment (ROI) due to the increases in efficiency they provide. Moreover, there is clear agreement that the cost–benefit of CPS is positive, both in environmental and financial terms. Developments in AI will further enhance the potential for operational improvements, vide infra.

Advanced planning and capital expenditure: The utility of cyber-physical systems goes well beyond improvements in operations. Within the digital twins, scenarios for the evolution of a system can be simulated—the scenario with the most advantageous cost-benefits for emission reduction is selected and implemented in the physical world. Here, machine intelligence will play a critical role; the digital twin is AI's playground, and on this playground AI can facilitate the determination of cost-optimal solutions for emission reduction. Consequently, in addition to the reduction in operational expenditure, capital expenditure will be reduced by avoiding e.g., over-engineering or superfluous infrastructure. This planning capability will lead to a systems evolution that gradually moves a certain system towards both lower costs and emissions while minimising the cost of this transition. Figure 33.1 illustrates both the current, operational improvements and the evolution of the system into the future. The reduction of capital expenditure will also reduce the carbon emissions embedded in a certain system. It can be concluded from this compendium that intelligent CPS are the ultimate decision support tool and will provide advanced asset management with optimised investment strategies. As the transition to net-zero emission will require significant financial resources, optimised investment strategies will be key for success.

Lifetime extensions: Yet another area in which intelligent CPS can provide essential benefits are asset-heavy industries such as manufacturing, chemicals or real estate. These industries are often locked into their infrastructure due to high capital costs and long lifetimes of assets; this lock-in creates a systems inertia that makes decarbonisation and meeting emission standards difficult. Here, CPS-facilitated improvements are the only economic route to decarbonisation and hence the improvements delivered by digital technology are pivotal for short- and medium-term emission targets. In addition, the avoidance of the decommissioning of still-intact infrastructure due to emission issues extends its lifetime and consequently increases the utility of the emissions embedded in the infrastructure (cradle-to-grave emissions). The scenario planning capability of the above-mentioned AI-supported digital twin will furthermore assist these industries in the avoidance of infrastructure lock-ins.

Feedback loops: As mentioned above, the electrification of the globalised economic system will create a complex, integrated system of sectors that either supply or demand energy. The beauty of this is that due to this integration, feedback loops are created that will facilitate the overall decarbonisation of human activity. Figure 33.2 depicts an illustrative example for photovoltaic cells (PVCs): The enhanced manufacturing of PVCs via cyber-physical production systems has reduced the resources and energy required for PVC production while increasing their efficiency. The former improvement reduces the carbon embedded in the cells and hence, the lifetime or cradle-to-grave emissions of electricity provision by PVC is lower. The latter improvement provides more electricity per area. Since electricity is also a key input factor for PVC production, a virtuous circle is created (Fig. 33.2).

CPS-enhanced grid management allows, moreover, for a higher percentage of intermittent renewables to be connected, which lowers the carbon footprint of electricity even more and another virtuous circle is formed. The same is true for battery electric vehicles that not only reduce tailpipe emissions, but also have the potential to provide distributed storage that smart grids can utilise to increase the share of renewables. CPS will be key to coordinating all electrified economic sectors while optimising the beneficial feedback loops outlined above. In addition, AI-enhanced CPS will create spillover effects: advantageous strategies developed for one sector will be applied to others as well. It can be concluded that CPS will be critical for the optimisation and integration of an electrified, low-carbon economic system.

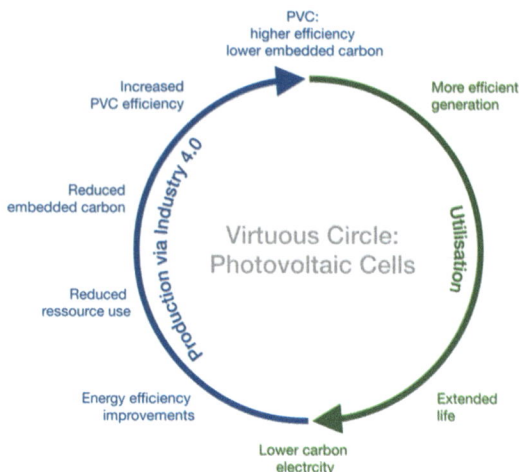

Fig. 33.2 Improvements driven by the Industry 4.0 concept create virtuous circles; many of the virtuous circles are connected and hence support other sectors

Information exchange and markets: Essential for all CPS is the efficient exchange of information. Dynamic knowledge graphs have proven to be an excellent facilitator that adds to the optimisation and planning capabilities of the digital twin within the CPS. In the knowledge graph, human knowledge is codified in ontologies which in turn are subject to algorithmic change by agents that are also part of the knowledge graph. This construction is a first step towards achieving AI 3.0 and shows already great potential with respect to solving the interoperability problem. Yet another area that needs to be improved is information asymmetry in markets e.g., for electricity or carbon. In this case, open data platforms could reduce asymmetry thereby enhancing the market forces that will help restructure economic activity to net-zero emissions. The proliferation of regimes that put a price on emissions, i.e., internalise the environmental externality, underline the importance of transparency and accountability— open data platforms and distributed ledger technology will play key roles in this market transformation.

Artificial intelligence: While artificial intelligence already has some influence in specific areas, mainly provided by deep learning, the benefits from its broad application will still take time to manifest. It can be deduced from the contributions in this compendium that the actual benefits of broad deployment of AI capabilities cannot yet be estimated or fathomed; in the past, AI developments surpassed even the most optimistic forecasts. While there are many reasons for this uncertainty, the key reason is the anticipated advent of quantum computing which is likely to trigger the development of artificial general intelligence and hence technological singularity. What the inflection point of singularity will bring is absolutely unforeseeable, but alas it raises the critical question of humanity's safety in the dawn of the era of superintelligence.

Risk mitigation: The potential arrival of a superintelligence raises many questions critical for humanity's survival: from this inflection point onwards, humans will not be the most intelligent species on this planet and machines with an IQ superior to that of humans may even develop some form of consciousness. How can it be ensured that a potential superintelligence serves humanity and the planet? This issue is referred to as the AI control problem and this compendium clearly argues that machine consciousness should be prevented by all means as long as the AI control problem is not solved. However, there are not yet any viable proposals for the solution of this conundrum and hence serious investments in AI safety research are called for to ensure humanity's safety and survival. With respect to CPS, there are also open questions, however, these are more practical and clearly solvable. Firstly, the governance system for technology has to be adapted and a combination of top-down and bottom-up governance is proposed herein to foster benefits and minimise risks. Secondly, legal systems have to be restructured so that they can keep up with the speed of technological developments; contributions propose agnostic and adaptive laws that include ethics-by-design. The capabilities of CPS, especially in monitoring and surveillance, could lead us to dystopian scenarios which have to be prevented at all costs. The importance that leaders assign to these developments is also clear to see from this compendium; significant investments, excellence centres and dedicated university courses are established to ensure that the benefits from this transition can be reaped while assuring that these advances serve humanity and the sustainability of our planet.

From this compendium it can be concluded that cyber-physical systems enhanced by artificial intelligence will be critical for the short-term reduction of emissions from economic activity and will moreover be a key tool to plan a fully decarbonised net-zero world. Thus, these technological developments will be pivotal in combating climate change and its potentially disastrous consequences while precautions must be implemented that ensure that the cure is not worse than the disease.

Recommendations for policy makers: From this compendium it becomes apparent that innovations like CPS and AI will provide benefits in all emission-intense sectors of the economy. Advances in the development of these digital technologies therefore will have ripple effects through the economic system and are likely to provide benefits in unforeseeable areas, both economic and environmental. The anticipated benefits of CPS and digital twin technology illustrate this—technologies that provide benefits in one sector spill over, i.e., they are implemented in related areas and provide similar benefits there.

Policy makers should therefore strategically support *both* the research and development. A portfolio of policy levers ranging from dedicated and targeted research grants and financial incentives for commercialisation to consortia and public private partnerships (PPP) should be applied to reap the benefits while reducing the risk the technologies pose. In addition, the transfer of academic knowledge via spin-offs should be enhanced using financial incentives. PPPs and consortia especially will foster the establishment of collective intelligence in this realm and such a collective intelligence is needed to tackle climate change.

The most important area, however, is AI safety research: here, proactivity is key—once an artificial general intelligence is established there is no way back and it should be known how to deal with such a machine intelligence before its inception. Governments should therefore invest significant amounts in AI safety research *now* to ensure the world is prepared for the most important and disruptive inflection point in human history, technological singularity.

Printed by Printforce, the Netherlands